THE LIFE & TIMES OF
CHUCKLES
THE
ROCKET DOG

A COMPANIONABLE GUIDE
TO POLYNOMIALS & QUADRATICS

by Linus Rollman, Greg Neps,
and ....

an Arbor Algebra book

The Arbor Center for Teaching
Tualatin, OR

Arbor Center for Teaching

This book was written, edited, and prepared for publication at Arbor School of Arts & Sciences, Tualatin, Oregon, during 2010 and 2011. The Arbor Algebra series is under the supervision of Kit Abel Hawkins, director of Arbor School, and Sarah Pope, publications editor for the Arbor Center for Teaching. Mary Elliott designed this book. Cover art, illustrations, and figures were drawn by Linus Rollman. The Arbor Algebra series is made possible by funding from the Bloomfield Family Foundation.

Rollman, Linus and Greg Neps.
    The Life & Times of Chuckles the Rocket Dog: A Companionable Guide
    to Polynomials & Quadratics.

(An Arbor Algebra book)
Summary: Volume III of a writing-based, common sense, whimsical & engaging introduction to algebra for middle-grade math students.

ISBN 978-0-9821363-5-5

Printed in the United States of America by Lightning Source, Inc.

TABLE OF CONTENTS

dedicated to the memory of Jack W. McKittrick,

patron and friend of the Arbor Center for Teaching

1

EXPONENTS

1 THE LAWS OF EXPONENTS

I'm going to start this book off with a chapter entirely devoted to exponents. Of course, you've been using exponents for quite some time now, but in this chapter I'm going to ask you to take a more in-depth look at the subject and learn some new things about it.

1. **Just as a quick review, what is the value of the expression 4^3 without an exponent?**

That was a simple one. I just wanted you to remind yourself what an exponent is and that 4^3 is definitely not the same thing as **4 · 3**.

The basic rule for how exponents work could be expressed like this:

$$x^a = \underbrace{x \cdot x \cdot x \cdot x \ldots}_{a \text{ times}}$$

And here's a piece of vocabulary that I'd like to remind you of: the power that a number is being raised to is the *exponent* and the number itself is called the *base*. So in the expression 4^3, **4** is the base and **3** is the exponent.

In this lesson, I'm going to ask you to figure out the Laws of Exponents. What, exactly, makes something a mathematical "law" is a subject that can be debated. It seems to me that the things that are referred to in this lesson as "laws" could just as easily be called rules. Calling them laws makes them sound more impressive, I suppose, and unchangeable. At any rate, there are five Laws of Exponents. You're already familiar with two of these Laws, though you haven't been calling them that up to this point.

2. **Simplify the expression x^5x^3.**
   (If you have any trouble with this, remember that, as a first step, you could write this expression as a long string of x's all multiplied by each other. For example, x^3 could be rewritten as x · x · x.)

Simplify the following expressions:

3. y^3y^7

4. m^2m^4

5. **In Problems 2 through 4 you applied the First Law of Exponents. Remember that when mathematicians write down rules (or laws), they like to write them in a generic, mathematical form. Here is the first part of the First Law of Exponents in its generic form. Complete the generic form, being sure to record the entire equation in your notebook:**

$x^a \cdot x^b =$

The First Law can be applied multiple times to a single expression. For example, you could use the First Law to rewrite $n^4m^3n^2m^8$ as n^6m^{11}.

Apply the First Law of Exponents to simplify the following expressions:

6. $y^3x^7y^2x^4$

7. $m^7m^3m^2n^4n^4n^3$

8. $(3a^4b^6)(7a^3bc^2)$
     (Don't let the presence of numbers as well as variables fool you here. In addition to applying the First Law, you should multiply 3 by 7.)

It can sometimes be useful to apply the First Law "backwards." The First Law states that $x^a \cdot x^b = x^{a+b}$, and an equal sign simply means that the two sides are the same, not that one side is the "problem" and the other is the "answer." So you could rewrite y^8 as y^6y^2 (or, of course, y^5y^3, or y^2y^5y, or any number of other possibilities).

Apply the First Law of Exponents to the following expressions. (There are multiple correct answers for each one.)

9. x^9

10. m^5

11. y^{23}

All right. That's the First Law of Exponents. Now it's time for the Second Law of Exponents. You've also used the Second Law many times before, particularly in the chapter in *Jousting Armadillos* called "Algebra & Fractions."

12. **Simplify the expression $\dfrac{y^6}{y^3}$.**

     (You can always consult your Note to Self book if you don't remember how to do this. Once again, it can also be helpful to think of an expression like y^2 as $y \cdot y$.)

Simplify the following expressions:

13. $\dfrac{m^5}{m^2}$

14. $\dfrac{x^{17}}{x^{10}}$

15. Here's the first part of the generic form of the Second Law of Exponents. Finish it (and write the whole equation in your notebook!):

     $\dfrac{x^a}{x^b} =$

Actually, to be technically correct, the Second Law of Exponents needs to be written like this:

$\frac{x^a}{x^b} = x^{a-b}$ as long as x is not equal to zero.

16. **Why do you have to include the part about the Law applying only when x is not equal to zero?**

Just like the First Law, the Second Law can be applied multiple times to simplify a single expression. (And you did this very thing many times in *Jousting Armadillos*.)

Apply the Second Law of Exponents to the following expressions:

17. $\dfrac{x^5 n^{10}}{xn^4}$

18. $\dfrac{3a^7 b^4}{5a^2 b^4}$

19. $\dfrac{75x^4 y^7 z^5}{15x^3 yz^3}$

In Problems 18 and 19, you had numbers to deal with along with variables. I hope you saw that in one case those numbers could be simplified and in the other case they couldn't.

Also like the First Law, the Second Law can be applied "backwards."

Use the Second Law of Exponents to rewrite the following expressions. (There are multiple — in fact, infinite — correct answers for each one.)

20. n^4

21. x^7

22. m^{20}

The First and Second Laws of Exponents can be applied to expressions in which the bases are numbers rather than variables. (This shouldn't come as a surprise — after all, the whole point of variables is that they can stand for numbers.) So, for example, $2^3 \cdot 2^5 = 2^8$. This ought to be true, since $2^3 \cdot 2^5$ means $(2 \cdot 2 \cdot 2)(2 \cdot 2 \cdot 2 \cdot 2 \cdot 2)$, which is, of course, 2^8.

For similar reasons, $\frac{2^5}{2^3} = 2^2$. However, you cannot apply the First Law to an expression like $3^4 \cdot 5^3$ or the Second Law to an expression like $\frac{7^6}{3^2}$.

23. **Explain why the First Law of Exponents doesn't apply to the expression $3^4 \cdot 5^3$.**

If you want to simplify an expression like $3^4 \cdot 5^3$, there really isn't anything to do other than calculate 3^4 and 5^3 and multiply them by each other.

Now for the Third Law of Exponents. It applies to expressions like this:

$(x^3)^4$

It won't be at all hard for you to figure out how to simplify an expression like that as long as you keep in mind what exponents actually mean.

24. In the case of the expression $(x^3)^4$, something is being raised to the fourth power. What is being raised to the fourth power?

25. If you were raising, say, y to the fourth power, you could write that out as $y \cdot y \cdot y \cdot y$. Rewrite $(x^3)^4$ in a similar way.

26. Based on the First Law of Exponents, you can rewrite the expression that you just wrote for Problem 25 as a single variable raised to a single power. Go ahead and do so.

Use what you figured out when you simplified $(x^3)^4$ to simplify the following expressions:

27. $(a^2)^4$

28. $(b^3)^3$

29. $(y^{15})^2$

30. $(3^4)^6$
   (As you see, the Third Law can be applied to numbers as well. You don't need to simplify this expression all the way to a single number. It would be very, very large! Three raised to a power will do just fine.)

31. Use the following to write a generic version of the Third Law of Exponents:

$(x^a)^b =$

As with the First and Second Laws, it might sometimes be useful to apply the Third Law "backwards." That is, you might want to rewrite x^{10} as $(x^5)^2$ or as $(x^2)^5$.

Use the Third Law of Exponents to find equivalent expressions for the following. (There are multiple correct answers for each problem.)

32. y^{15}

33. a^{12}

34. 4^{20}

35. Why is it difficult to rewrite x^{13} in a similar way?
   (Hint: It's actually not that difficult — it's just kind of silly.)

As you've seen already, the Laws of Exponents are pretty easy to figure out and work with as long as you keep in mind what exponents mean and how they work. The same thing is true for the Fourth and Fifth Laws.

The Fourth Law applies to a situation like this:

$(xy)^3$

36. In the case of $(xy)^3$, what are you raising to the third power?

37. a^3 could be rewritten a · a · a. How could $(xy)^3$ be rewritten?

38. As you know perfectly well, a string of things multiplied together can be multiplied in any order, so rewrite your string of xy's to put all of the x's together and all of the y's together. Then rewrite that expression as x to a power multiplied by y to a power.

That result almost certainly came as no surprise to you. Use what you've just figured out to simplify the following expressions:

39. $(ab)^4$

40. $(2a)^3$ (When you're dealing with numbers instead of variables, it's really up to you to decide what form is simplest. In this case, do you prefer 2^3 or 8?)

41. $(7x)^5$

42. $(xyz)^4$

43. Write a generic form of the Fourth Law of Exponents:

$(xy)^a =$

Combine the Third and the Fourth Laws to simplify the following expressions:

44. $(x^3y^3)^4$

45. $(a^2b^3)^5$

46. $(3m^4y^{10})^3$

Just as you did with the first three Laws, you ought to be able to apply the Fourth Law "backwards."

Rewrite the following expressions using the Fourth Law. (Sometimes you'll be applying the Third Law as well, and yes, there are multiple correct answers.)

47. a^7b^7 48. $a^{14}b^{21}$

49. $a^6b^3c^{12}$ 50. $27x^3$

All right, one more Law of Exponents to deal with. The Fifth Law is really just the division version of the Fourth Law. This time I'll go ahead and give you the generic version and let you apply it.

The Fifth Law of Exponents:

$$\left(\frac{x}{y}\right)^a = \frac{x^a}{y^a}$$ as long as y is not equal to zero.

51. Once again, why do I have to specify that y can't be equal to zero? (For reasons that you will find out in the next lesson, it's perfectly fine if a is equal to zero.)

Simplify the following expressions according to the Fifth Law:

52. $\left(\dfrac{a}{b}\right)^4$

53. $\left(\dfrac{3}{y^2}\right)^3$ (On this one you'll be using the Third Law as well as the Fifth.)

Of course, you should be able to go "backwards" as well. (Yup, there are multiple correct answers.)

54. $\dfrac{a^{10}}{b^5}$

55. $\dfrac{x^4}{16}$

56. Now would be a good time to write a **Note to Self** about the **Laws of Exponents**. Really all you need in this Note is the generic version of each Law and a couple of examples of how each one is applied. This seems like a good time to mention that I don't expect you to memorize which Law is which — as in, I'd never ask you, "Quick! What's the Third Law of Exponents?" (Although it's possible that your teacher might disagree with me about this. You should check with him or her.) I do expect you to know how all of the Laws work and to apply them to the appropriate sorts of expressions.

I'm going to give you a few more problems of a slightly different type that involve applying the Laws. Take the First Law, for example. It states that $x^a \cdot x^b = x^{a+b}$. As I've said already, you can replace the x with a number and get a true statement, as in $2^3 \cdot 2^5 = 2^8$. But a variable can also stand for an expression. So, in other words, the First Law could be applied like this: $(y + 4)^3 \cdot (y + 4)^5 = (y + 4)^8$. All I've done there is to replace the x in the First Law with the expression $(y + 4)$, which is a perfectly legitimate move.

In the following problems, use the Laws of Exponents to simplify the expressions. However, your simplified versions should still contain parentheses. Later in this book, you'll spend quite a bit of time simplifying expressions like, for example, $(x + 3)^2$ so that they no longer have parentheses, but for now, leave the parentheses.

57. $(x + 5)^3 \cdot (x + 5)^5$

58. $\dfrac{(b - 5)^7}{(b - 5)^4}$

59. $(y - 3)^4 \cdot (y - 3)^2$

60. $((a + 3)^3 (b - 2)^4)^{10}$

61. $\left(\dfrac{(x + y)^5}{(x + y)^3} \right)^3$

62. Explain why the First Law of Exponents does not apply in this situation:

$(y + 4)^3 \cdot (y + 6)^5$

REVIEW

Convert the following equations into two-intercept form and state their x- and y-intercepts:

1. $y = \dfrac{3}{4}x + 3$

2. $y = -\dfrac{3}{5}x - 3$

3. Graph the equation $y = (x - 2)^2 + 4$.

4. Solve the equation $\dfrac{x - 2}{2} = \dfrac{2x + 2}{5} + 1$.

Write equations to go with the following tables and tell what the basic shape of the graph would be (straight line, parabola, or hyperbola):

5.

x	-2	-1	0	1	2
y	16	13	10	7	4

6.

x	0	1	2	3	4	5	6
y	9	4	1	0	1	4	9

7.

x	-6	-4	-2	0	2	4	6
y	20	10	4	2	4	10	20

8. John Carter can paint a palace room in 3 hours. It takes Dejah Thoris 7 hours to paint a palace room. (She's considerably more careful.) How long does it take them to paint such a room working together (assuming that no one is bothered by John Carter's sloppiness)?

9. Humans, as you may know, have four limbs apiece. They also (at least in this story) carry one sword apiece. On the other hand, the green men of Mars have six limbs apiece and carry two swords apiece. In a mixed gathering at which there are 76 limbs and 22 swords, how many of each type of being are there?

10. Here are a couple of logic puzzles based on ones by Norman Willis:

I've just given you two boxes. One contains a chocolate éclair and the other contains a rabid weasel. The labels on the boxes read as follows:

> Only one of these
> labels is false.

> You should open
> this box. Yum!

The rabid weasel is currently asleep, so you won't hear any deranged scuffling, and I'm not going to let you pick up the boxes and shake them. If you must open one of the boxes, which should it be and why?

Assuming you survived the last two boxes, I'm going to give you two more, one box containing a gift certificate to Dr. Enid's House of Extravagant False Moustaches and the other containing an entry ticket to Carl Denham's House of Angry Free-Range Gorillas. The boxes are labeled as follows:

> These labels are
> both false.

> Open this box if
> you ♥ moustaches.

If you must open a box and use its contents, which box should it be and why? And am I a good friend or what?

2 NEGATIVE EXPONENTS

Yes, there are definitely such things as negative exponents, though you haven't yet encountered them (at least not in this series of textbooks). They are actually pretty easy to deal with once you get over one perhaps rather surprising fact: negative exponents are almost nothing like negative numbers.

Negative numbers, as you well know, are numbers that are below zero. For this reason, when people first encounter, say, 2^{-2}, their gut instinct is generally to say something like, "Oh! 2^{-2} is probably equal to **-4**." This would not be an unreasonable guess, based on what you know about negative numbers and exponents, *but it's not true*. Generally, I try to avoid writing things that aren't true, just in case someone is skimming through these pages, so let me be perfectly clear:

2^{-2} is NOT equal to -4!

This is one of those things (kind of like the divide-by-zero rule) where I've seen quite a few students over the years learn about negative exponents, be perfectly convinced that 2^{-2} is not equal to **-4**, and then, after they haven't used negative exponents for a while, they'll come across one and — once again — they'll say something like, "Oh! 2^{-2}. That must be equal to **-4**."

1. **In your notebook, boldly and clearly, I want you to write the statement, "2^{-2} is NOT equal to -4." Now I want you to say it aloud ten times: "Two to the power of negative two is not equal to negative four!" Really. Say it out loud, like you mean it. Good. Now, you promise to remember that, right?**

Before I get into how negative exponents *do* work, I'm going to take a minor detour to talk about the following expression:

$$-2^2$$

You could debate for quite awhile about whether that expression is equal to **4** or to **-4**. There's a good case to be made for both. It should be equal to **4** if -2^2 means **(-2)(-2)**, because a negative times a negative is a positive. On the other hand, it should be equal to **-4** if -2^2 means **-(2)(2)**, because in that case you'd be squaring the two first and then making the result negative. I think either way makes sense. As it happens, mathematicians have agreed that -2^2 means **-(2)(2)**, or, in other words:

$$-2^2 = -4$$

When mathematicians want to use exponents to indicate that negative two should be multiplied by itself, they write it **(-2)²**. So:

$$(-2)^2 = 4$$

For the following expressions, decide whether the final, simplified version would be negative or positive. (Notice that I'm not actually asking you to do any calculating.)

2. -6^2

3. $(-9)^2$

4. -5^3

5. $(-5)^3$ (Remember how multiplying negatives works!)

6. $(2 - 3)^4$

7. -5^6

8. $(-4)(-15)^4$

9. $((-8)(11))^6$

10. $((4)(-3))^{13}$

Thanks for taking that little detour. It was important, even though it didn't really have to do with the main subject, which is negative exponents. Before you figure out how negative exponents work, here's a quick review question:

11. $2^{-2} = -4$, right?

Okay. Now you're ready. In order to understand what negative exponents mean, you need to be sure that you understand this pattern:

4   8   16   32   64   128   ...

That, of course, is simply the increasing powers of two, written in order: $2^2 = 4$, $2^3 = 8$, $2^4 = 16$, and so on.

12. **What happens to the numbers in the pattern with each increasing step?**

4   8   16   32   64   128   ...

That makes perfect sense, right? Each step is another power of two, which simply means multiply by another two. So, following that same logic...

13. **What happens to the numbers in the pattern with each step moving in the opposite direction, from larger to smaller?**

128   64   32   16   8   ...

Understanding negative exponents (as well as the exponents one and zero) just means following the logic of that pattern a little further.

14. **If 2^3 is 8 and 2^2 is 4, what's 2^1? (You actually know this one already.)**

15. **Here's a new concept. Following that exact same pattern, what must 2^0 be?**

16. **And now for the negative exponents. Continuing that pattern one more step, what's 2^{-1}?**

I hope you find that pattern convincing. Negative exponents don't really have to do with negative numbers; they have to do with fractions. As you'll shortly see, not only does this make sense in terms of the pattern of doubling or halving, but it also means that negative exponents fit in perfectly and intelligibly with the Laws of Exponents.

17. **Copy the chart below into your notebook and finish filling it out, following the pattern you've established:**

2^5	2^4	2^3	2^2	2^1	2^0	2^{-1}	2^{-2}	2^{-3}	2^{-4}	2^{-5}
32	16									

Realize that your chart could be extended infinitely in either direction, with the powers of two becoming huger and huger in one direction and tinier and tinier in the other direction.

Copy and finish filling out the following charts:

18.

3^5	3^4	3^3	3^2	3^1	3^0	3^{-1}	3^{-2}	3^{-3}	3^{-4}	3^{-5}
243	81									

19.

4^5	4^4	4^3	4^2	4^1	4^0	4^{-1}	4^{-2}	4^{-3}	4^{-4}	4^{-5}
1,024	256									

20. **Based on those examples, what happens to a number when you raise it to the power of one?**

21. **Based on those examples, what is any number raised to the power of zero equal to?**

22. **Based on those examples, what is the relationship between a number raised to a positive power and the same number raised to the same negative power, such as 2^3 and 2^{-3} or 4^2 and 4^{-2} or 5^4 and 5^{-4}?**

The relationship that you discovered through those charts and described in Problem 22 can be summed up in the following generic rule for negative exponents:

$$x^{-a} = \frac{1}{x^a}$$

(Or, to put it differently, the two numbers are reciprocals of one another.)

And the other two rules that you just figured out can be summed up in this way:

$$x^1 = x$$
$$x^0 = 1$$

And to sum up the whole concept of negative exponents in one simple way: *just as increasing exponents indicate repeated multiplication, decreasing exponents (on into the negatives) indicate repeated division.*

Rewrite the following expressions according to the rules you discovered. You can choose whether the rewritten expressions contain exponents or not. (In other words, $\frac{1}{9^8}$ is a valid way of rewriting 9^{-8}.)

23. 7^{-3}

24. x^{-4}

25. 30^1

26. $2,367,498^0$

State whether the simplified versions of the following expressions will be greater than, equal to, or less than one:

27. $15 \cdot 5^{-2}$ (Notice that the exponent applies only to the 5 and remember your Order of Operations (PEMDAS)).

28. $9(2)^{-3}$

29. $(99^4)(99^{-4})$

It's interesting to note that, as long as the base of an expression is positive, no matter how negative the exponent is, *you never get a negative number.* $3^{-1,000,000}$ is an *extremely* small number (it's one split into three parts, each part split into three parts, each part split into three parts — a million times!), but it's still greater than zero.

Now, if the base itself is negative, you can get a negative number when you raise it to a negative power. This works in exactly the same way as the detour problems at the beginning of this lesson. As you now know:

$$2^{-2} = \frac{1}{2 \cdot 2} = \frac{1}{4}$$

Well, following the rule from earlier:

$$-2^{-2} = -\left(\frac{1}{2 \cdot 2}\right) = -\frac{1}{4}$$

... and:

$$(-2)^{-2} = \frac{1}{(-2)(-2)} = \frac{1}{4}$$

State whether the simplified versions of the following expressions will be positive or negative:

30. -3^{-2}

31. $(-3)^{-2}$

32. -6^{-5}

33. $(-6)^{-5}$

State whether the simplified versions will be greater than one, equal to one, between one and zero, between zero and negative one, equal to negative one, or less than negative one:

34. $(-4)^{-2}$

35. $20(-4)^{-2}$

36. $(-5)^{-4}(-3)^3$

37. $10(-2)^{-3}$

38. $(-75)^3 \cdot (-75)^{-3}$

As I said earlier, negative exponents are perfectly compatible with the Laws of Exponents.

39. Take the case of this expression: $x^4 \cdot x^{-2}$. You can rewrite x^4 as $x \cdot x \cdot x \cdot x$. Rewrite x^{-2} in a similar fashion. Then multiply those two new versions of x^4 and x^{-2} by each other and simplify the result.

The First Law of Exponents states:

$x^a x^b = x^{a+b}$

40. Multiply x^4 by x^{-2} according to the First Law of Exponents. (I'm sure I don't need to remind you what happens when you add a positive and a negative number.)

41. Now consider the expression $\dfrac{x^5}{x^{-3}}$.

Rewrite this expression using the expanded versions of x^5 and x^{-3} (that is, $x \cdot x \cdot x \cdot x \cdot x$ and $\dfrac{1}{x \cdot x \cdot x}$).

Now do that division problem (I will remind you here that in order to divide by a fraction, you multiply by its reciprocal) and simplify the result.

The Second Law of Exponents states:

$\dfrac{x^a}{x^b} = x^{a-b}$ as long as x is not equal to zero.

42. Use the Second Law of Exponents to simplify the expression $\dfrac{x^5}{x^{-3}}$.

43. Now for the Third Law. Use the expanded version of x^3 — just as you did in Problems 39 and 41 — to simplify the expression $(x^3)^{-2}$.

The Third Law of Exponents tells us:

$(x^a)^b = x^{a \cdot b}$

44. Use the Third Law to simplify $(x^3)^{-2}$.

I'm going to assume that it's pretty clear that the Fourth and Fifth Laws apply to expressions with negative exponents without having you prove it. Remember that the Fourth Law is:

$$(xy)^a = x^a \cdot y^a$$

… and the Fifth is:

$$\left(\frac{x}{y}\right)^a = \frac{x^a}{y^a}$$ as long as y is not equal to zero.

Simplify the following expressions using whatever combination of the Laws of Exponents is appropriate for each case:

45. $(3^2)(3^{-1})$

46. $\dfrac{y^3}{y^5}$

47. $\dfrac{m^5}{m^{-5}}$

48. $(3^3)(3^2)(3^1)(3^0)(3^{-1})(3^{-2})(3^{-3})$

49. $(3x)^{-3}$

50. $\dfrac{n^{-4}}{n^{-6}}$

51. $\dfrac{x^7 y^{-3}}{x^{-2} y^2}$

52. $\left(\dfrac{3^2}{3^{-4}}\right)^2$

53. $\left(\dfrac{3^2}{3^{-4}}\right)^{-2}$

54. $(-x^3)^{-4}$

55. $(-x^{-3})^4$

56. $\left(\dfrac{x^2 y^{-3}}{x^{-5} y^4}\right)^{-2}$

57. $\left(\dfrac{x}{y}\right)^{-1}$

Before this lesson ends, I want to give you a reminder about a subject that I've mentioned several times before. In that entire previous problem set, all you were really doing was applying the Laws of Exponents in order to change the forms of various expressions. You were not finding the "answers" to a series of problems. Take Problem 57, for example. There were a number of perfectly legitimate ways to rewrite that expression. You could have written it as $\dfrac{x^{-1}}{y^{-1}}$ or as $x^{-1}y^1$ or as $x^{-1}y$ or as $\dfrac{y}{x}$ — and there are some other

possibilities as well — because *those expressions all mean the same thing: they're all equal to one another*. And none of them is necessarily any better than the original expression, $\left(\dfrac{x}{y}\right)^{-1}$.

One of the most important skills in algebra is simply being able to change expressions from one form to another — that is, being able to see when one thing is equal to another.

58. It's time for a *Note to Self* about *negative exponents and zero as an exponent*. Probably the most important thing to include is an explanation of what negative exponents and a zero exponent mean. I would suggest also choosing a few problems from this lesson that involve applying the Laws of Exponents to expressions with negative exponents.

59. One final question. $2^{-2} = -4$, doesn't it?

REVIEW

State the shared solutions of the following pairs of equations:

1. $y - 3x = -5$
    $y = x + 7$

2. $2y - 3x = -30$
    $3y + 2x = -19$

3. $2y - 6x = -10$
    $y = 3x + 7$

4. Graph the inequality $6 - 2y \geq 10x$.

5. Find the greatest common factor of $315n^3m$ and $30nm^3$.

6. Find the least common multiple of $440x^2y$ and $350xy^2$.

7. Find the value of the expression $\dfrac{3x + 3y}{7(2x - y)}$ when $x = 4$ and $y = 8$.

8. Last year, the newt population in Gussie Fink-Nottle's pond grew by 6%. If there were 265 newts at the end of the year, how many were there at the beginning of the year?

9. Mr. Toad was driving his motorcar through the countryside. He traveled from Bentley-on-Thames to Snodgrass-on-Tweed at 50 miles per hour and stopped for lunch. Then he drove to Bloodpudding-on-Toast at 60 miles per hour. If he drove a total of 10 hours and traveled a total of 565 miles, how long did he spend traveling between each of the quaint villages?

10. This puzzle is based on one by Paul Bostock and Elliott Line:

    The people of Kneecapsia have a curious (and limited) economy. Items in their market-place cost anywhere from 1 to 15 patelli — Kneecapsicans can't actually count any higher than 15 — and you must always pay with exact change, because they don't understand the principle of subtraction. No merchant will accept more than three coins for any one purchase and there are only three denominations of coin. What are those three denominations?

3 SCIENTIFIC NOTATION

Back in Chapter 2, Lesson 5 of *Jousting Armadillos*, I asked you to look at what happens when you raise **10** to a power and I told you that what you discovered would come in very handy later on when you studied scientific notation. Well, "later on" is now.

1. Rewrite the expression 10^1 without an exponent.

2. Rewrite the expression 10^2 without an exponent.

3. Rewrite the expression 10^3 without an exponent.

4. Do the same for 10^4, 10^5, and 10^6.

5. **Without writing it out, how many zeros would you expect to follow the 1 if the expression 10^{25} were written without an exponent? How about 10^{75}? 10^{100}?**

If you look up "names of large numbers" online, you will find that there are amusing names for some outrageously large numbers. For instance, according to the internet, 10^{213} is called "one septuagintillion" (or, in French, "quinquatrigintilliard"), and 10^{453} has the improbable name of "one quinquagintacentillion." As far as I know, no one ever actually uses those names, but I think it's kind of wonderful that they exist.

6. **You already know the names for 10^3, 10^6, and 10^9, among others. What are those names?**

So raising **10** to a power gives you a very convenient way of writing out a number like one quinquagintacentillion: you can write 10^{453} instead of a **1** followed by **453** zeros. Now let's look at the next step in scientific notation.

7. **Consider the expression $2 \cdot 10^9$. First of all, write down an English translation of that expression. (In Problem 6 you wrote the name for one 10^9 — so what's the name for two of them? That's what multiplying something by two means, after all.) Next, rewrite the expression as a single number without a multiplication sign or an exponent.**

8. **Write the English names for the following expressions and rewrite them without multiplication signs or exponents: $7 \cdot 10^3$, $5 \cdot 10^5$, $9 \cdot 10^6$.**

9. **Write the expressions $2 \cdot 10^{10}$, $2 \cdot 10^{11}$, and $2 \cdot 10^{12}$, each without multiplication symbols or exponents. How many *times* larger is $2 \cdot 10^{10}$ than $2 \cdot 10^9$? How much *greater* (as in, if one is subtracted from the other) is $2 \cdot 10^{10}$ than $2 \cdot 10^9$? How many *times* larger is $2 \cdot 10^{11}$ than $2 \cdot 10^{10}$? How much *greater* (subtraction style) is $2 \cdot 10^{11}$ than $2 \cdot 10^{10}$? How many *times* larger is $2 \cdot 10^{12}$ than $2 \cdot 10^{11}$? How many *times* larger is $2 \cdot 10^{12}$ than $2 \cdot 10^9$?**

10. Now I'll make things slightly trickier. Consider the expression $2.1 \cdot 10^9$. (This expression is usually read, "two point one times ten to the ninth," even though "two and one-tenth" is technically better than "two point one.") I imagine it was pretty easy for you to rewrite $2 \cdot 10^9$ without a multiplication sign or an exponent. I'm going to ask you to do the same thing for $2.1 \cdot 10^9$. This may seem more difficult, except that multiplying a decimal by ten, or by any multiple of ten, is *extremely easy*. I am not going to remind you how to do it — you can figure it out! (Here's one hint: Some people want to say that $2.1 \cdot 10^9$ is equal to 2,000,000,000.1 and it's definitely not.)

11. How much greater is $2.1 \cdot 10^9$ than $2 \cdot 10^9$?

12. 2,100,000,000 is the "decimal form" of $2.1 \cdot 10^9$. Without writing out the decimal forms, how many times larger is $2.1 \cdot 10^{11}$ than $2.1 \cdot 10^{10}$?

13. Without writing out the decimal forms, how much bigger is $4.2 \cdot 10^{12}$ than $4 \cdot 10^{12}$? (You can write out the decimal forms if you really need to, but try doing it without. The real question, on some level, is how big is the ".2" part of $4.2 \cdot 10^{12}$?)

14. Write the decimal form of $2.13 \cdot 10^9$. If I had, say, $2.13 \cdot 10^9$ porcupines (lucky me!), how many porcupines would the "3" in $2.13 \cdot 10^9$ stand for?

Convert the following numbers to decimal notation:

15. $7.8 \cdot 10^7$ 16. $3.45 \cdot 10^8$

17. $5.873 \cdot 10^5$ 18. $6.00034 \cdot 10^{10}$

19. $9.0203776 \cdot 10^{14}$

So, when you write the number **5,230,000** in the form **$5.23 \cdot 10^6$**, you're using scientific notation. It's called "scientific notation" because, well, it's used very frequently by scientists. It's probably already becoming clear to you why scientists might like to use scientific notation, but I'll give you some specific examples of how it's useful.

When you study chemistry, one of the things that you'll do is keep track of chemical reactions. (It's a branch of chemistry called *stoichiometry*.) Those reactions happen between very small pieces of substances called *molecules*, and one of your jobs will be to keep track of how many molecules take part in a reaction. When chemists count molecules, they use a particular unit, just like counting eggs by the dozen. This unit is named, funnily enough, a *mole*. (And that's pronounced just like the animal, not like the beginning of *molecule*.) Here, in a slightly rounded version, is how many molecules are in a mole:

602,200,000,000,000,000,000,000

(To give you an idea of how tiny a molecule is, one mole of carbon weighs twelve grams: less than one-tenth the weight of a regulation baseball.) In scientific notation, the number of molecules in a mole looks like this:

$6.022 \cdot 10^{23}$

Suppose that you decide you prefer to study astronomy rather than chemistry. In that case, you might find that the distance between Earth and the Canis Major Dwarf Galaxy (the closest galaxy to our own) is:

236,000,000,000,000,000 km

Or, in scientific notation:

$2.36 \cdot 10^{17}$ km

If you choose to study biology, you might be interested to know that the number of red blood cells in an adult male human's body is (roughly):

30,000,000,000,000

Or, in scientific notation:

$3 \cdot 10^{13}$

20. **Just based on looking at the two versions of the number of molecules in a mole, or the distance between our galaxy and its nearest neighbor, or the number of red blood cells in a man's body, why do you think that a reasonable person might prefer to use scientific notation rather than decimal notation in these cases?**

In all honesty, the answer is no more complicated than that: numbers in scientific notation are shorter and easier to write. If you were a scientist, or for that matter a textbook writer or publisher, you'd prefer to write $6.022 \cdot 10^{23}$ over and over again rather than 602,200,000,000,000,000,000,000. It's also easier to understand what "six point zero two two times ten to the twenty-third" means, rather than "six hundred and two sextillion, two hundred quintillion," which is the other option. Plus, as you will see shortly, numbers in scientific notation are easy to do some kinds of calculations with because you know the Laws of Exponents.

Just as scientific notation gives us a convenient way of writing and dealing with very large numbers, it also helps in dealing with very small numbers. For instance, if we go back to those molecules in a mole, you know that if $6.022 \cdot 10^{23}$ of them can weigh a lot less than a baseball, they must be tiny. In fact, one tiny atom of carbon (because carbon is an element, it actually consists of atoms rather than molecules, but don't worry about that now) is somewhere around **30 trillionths** of a meter in diameter. Scientific notation can help us cope with such a small number. Scientific notation involves exponents, and something like a trillionth of a meter involves dividing a meter into tiny parts. Well, you know a bit about using exponents to represent dividing something up...

21. **Write the decimal forms of $2 \cdot 10^3$, $2 \cdot 10^2$, $2 \cdot 10^1$, and $2 \cdot 10^0$. (Don't let that last one trick you — you know perfectly well what you get when you raise anything to the power of zero.) What fraction of $2 \cdot 10^3$ is $2 \cdot 10^2$? What fraction of $2 \cdot 10^2$ is $2 \cdot 10^1$? What fraction of $2 \cdot 10^1$ is $2 \cdot 10^0$?**

22. **Following the pattern that you just found, what fraction of $2 \cdot 10^0$ must $2 \cdot 10^{-1}$ be? Write the decimal form of $2 \cdot 10^{-1}$.**

23. Using similar logic, write the decimal forms $2 \cdot 10^{-2}$, $2 \cdot 10^{-3}$, and $2 \cdot 10^{-4}$.

24. Based on the work that you've just done, what is a simple rule for converting from scientific notation with a negative exponent to decimal form? (That is, based on the exponent, how far do you move the decimal point to the left?)

25. As I mentioned above, the radius of a carbon atom is roughly 30 trillionths of a meter. That's $30 \cdot 10^{-12}$ meters. Write $30 \cdot 10^{-12}$ in decimal form.

Before I give you a few numbers to practice converting from decimal notation into scientific notation, I need to say a few words about "official" scientific notation. But first, let me ask you a question.

26. How many times bigger is $1.37 \cdot 10^{15}$ than $13.7 \cdot 10^{14}$?

As you know, I'm not above throwing in the occasional trick question if I think it's for a good purpose. As I hope you were able to see, $1.37 \cdot 10^{15}$ is, in fact, the same size as $13.7 \cdot 10^{14}$. They're just two ways of writing exactly the same number. That same number could also be written as $.137 \cdot 10^{16}$, or in an infinite number of other ways. But the only way of writing it that counts as "official" scientific notation is $1.37 \cdot 10^{15}$. A number in "official" scientific notation has only one digit in front of the decimal place.

It's probably clear from my use of quotation marks around the word *official* in the last paragraph, but I personally think that this is a little bit silly. As far as I'm concerned, $1.37 \cdot 10^{15}$ and $13.7 \cdot 10^{14}$ are both pretty easy and efficient ways of writing the same number, and they're both definitely easier than the decimal form: 1,370,000,000,000,000. So I'd say that you should go ahead and use either one as long as it makes your work easier to do and to understand. However, your teacher might disagree with me, and there's a pretty decent possibility that at some time in the future, on a quiz in a science class, you'll be asked to use scientific notation and you'll lose points if you write $13.7 \cdot 10^{14}$ instead of $1.37 \cdot 10^{15}$ and you'll blame me because I never told you and you'll harbor bitterness in your heart about it forever. So, to save both of us that pain, let me repeat that *in "official" scientific notation, there should be only one digit in front of the decimal point.* (Also, in "official" scientific notation, there should always *be* a decimal point, so you should write $2.0 \cdot 10^{10}$ instead of $2 \cdot 10^{10}$.)

Write the following numbers in scientific notation and — what the heck, it'll be good practice — make sure it's "official" scientific notation:

27. 8,259,000,000

28. 32,623,000,000,000

29. 5,438,200,000,000,000,000

30. 0.00325

31. 0.0000450032

32. 0.0000000000005

33. 1

One good reason to use scientific notation is that it saves ink and paper and time by allowing you to write very large or very small numbers in an efficient way. But another reason is that it greatly simplifies certain kinds of calculations with very small or very large numbers.

Remember that an average carbon atom is **30-trillionths** of a meter across — in scientific notation, that would be **3.0 • 10^{-11}** meters. Suppose you were to string together, say, **4 billion** carbon atoms in a row and you wanted to know how long the string was. One way of doing that would be to write out both numbers in decimal notation and do the calculation. Well, **4 billion** is **4,000,000,000** and **3.0 • 10^{-11}** is **.00000000003**. Great, all I have to do is multiply **3** by **4** and then move the decimal point around. So let's see, I have to move it to the right nine times and then back to the left ten times — no, eleven times — let me count it again… oh wait, this is a pain.

34. **In scientific notation, 4,000,000,000 is 4 • 10^9. So your job is to multiply 3.0 • 10^{-11} by 4.0 • 10^9. There are a couple of things that make that very easy to do. The first is the Associative Rule of Multiplication (look it up in your Note to Self book from last year if you like) which tells you that when you're multiplying a string of things like 3.0 • 10^{-11} • 4.0 • 10^9, you can do that multiplication in any order. So do it like this instead: (3.0 • 4.0)(10^{-11} • 10^9). The three times four part is pretty easy, right? Write down the answer. But the 10^{-11} • 10^9 part is easy, too, since you know the Laws of Exponents. What is 10^{-11} • 10^9? And therefore, what's the answer to the multiplication problem? (It should look like a number times ten to a power — in other words, a number in scientific notation.)**

I hope that the answer you got to the last problem was **12.0 • 10^{-2}**. That's how long the string of carbon atoms would be in meters. If you prefer decimal form, that's **0.12** meters long, or **12** centimeters. (Just think about that for a second: a string of **4 billion** of these things is **12** centimeters long!)

Let's suppose you wanted to calculate how long it would take to travel to the Canis Major Dwarf Galaxy, **2.36 • 10^{17}** km away, if you could travel the speed of light, which is approximately **2.6 • 10^{11}** km per day. You'd set up your division problem like this:

$$\frac{2.36 \cdot 10^{17}}{2.6 \cdot 10^{11}}$$

Think of that as two division problems (or two fractions, if you prefer) multiplied by each other:

$$\left(\frac{2.36}{2.6}\right)\left(\frac{10^{17}}{10^{11}}\right)$$

My trusty calculator tells me that **2.36** divided by **2.6** is roughly **0.9**.

35. **Using the Laws of Exponents, what's 10^{17} divided by 10^{11} (expressed as a power of ten, of course)? Therefore, in scientific notation, how many days would it take to travel from here to the Canis Major Dwarf Galaxy at the speed of light?**

If you got what I got, it would take about $0.9 \cdot 10^6$ days (or $9.0 \cdot 10^5$). By the way, that's about **2,500** years, in case anyone's counting. Clearly we have more than a few obstacles to overcome before this kind of space travel will be possible.

Solve the following problems. Go ahead and express your answers in "official" scientific notation. Ask your teacher whether you can use a calculator for the decimal multiplication and division. (None of the calculations are too hard.)

36. $(3.2 \cdot 10^{12})(4.1 \cdot 10^{14})$

37. $(4.03 \cdot 10^{10})(5.0 \cdot 10^{-20})$

38. $\dfrac{7.04 \cdot 10^1}{2.2 \cdot 10^4}$

39. $\dfrac{2.345 \cdot 10^{20}}{6.7 \cdot 10^{13}}$

40. $\dfrac{7.0 \cdot 10^5}{2.5 \cdot 10^{-50}}$

Solve the following problems. It may sometimes be necessary to convert numbers into scientific notation to do so.

41. If I had $2.13 \cdot 10^9$ porcupines and each porcupine had 30,000 quills, how many quills would I have?

42. If the average soccer field has approximately $8.2 \cdot 10^{10}$ blades of grass, how many blades of grass would there be on 100 soccer fields?

43. The weight of the earth is approximately $1.3176 \cdot 10^7$ pounds. The weight of a pangolin (an anteater-like creature covered with plate-like scales) is about 20 pounds. How many pangolins would you need to equal the weight of the earth?

44. The distance from my house in Portland, Oregon, USA to Mumbai India is about 8,000 miles. The distance from my house the sun is about 93 million miles. How many times greater is the distance from my house to the sun than from my house to Mumbai?

45. It's time to write a **Note to Self** about **scientific notation**. It should use examples to show how scientific notation can be used to write very large and very small numbers and how to multiply and divide numbers in scientific notation.

As you can see, multiplying and dividing numbers in scientific notation is reasonable easy and potentially useful. You might have reason to hope that the same thing would be true for addition and subtraction, but, sadly, it isn't. There really is no shortcut for adding and subtracting numbers in scientific notation — in fact, quite often the best thing to do is to convert them to decimal notation first.

1. Graph the shared solutions of the following set of inequalities:

    $y + 2x > 3$

    $3x - y \geq 2$

2. What is the slope of a straight line that passes through the points (6, 15) and (0, -3)?

3. Solve the following inequality and represent its solution on a number line:

    $-2x - 6 < 9 + x$

4. Simplify the expression $\frac{2x^2}{7} - \frac{x^2}{3}$.

5. A giant inflatable moose is built on the scale of 15 : 2. If the inflatable antlers are 26 ¼ feet across, how broad are the real antlers? (Please set up a proportion to solve this.)

6. Greg really likes getting compliments. If Linus compliments Greg ten times a day, Greg is happy and will leave him alone. But if Linus doesn't say anything nice for the whole day, Greg will take 6 dollars from him. (In real life, Greg is a very nice guy. That counts as one compliment.) Assuming there's a linear relationship between the compliments and the extortion, write an equation in two-intercept form where y is the amount of money Greg takes from Linus and x is the number of compliments Linus gives Greg in a day.

7. What inequality is represented by this graph?

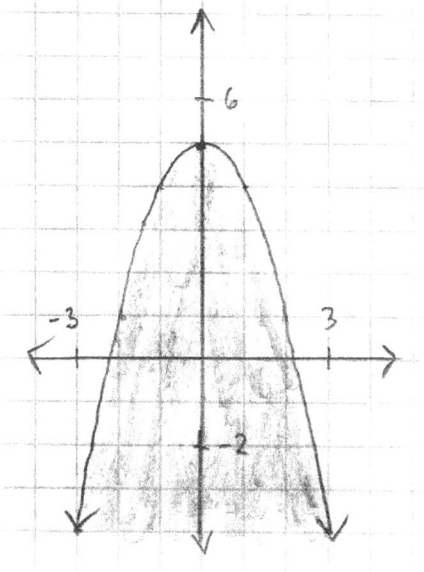

8. Angus owns 75 chickens; some are Fluffy-Neck chickens, some are Naked-Neck chickens, and some are No-Neck chickens. The ratio of Fluffy-Necks to Naked-Necks is 7 : 3. There are 10 more No-Necks than Naked-Necks. How many of each variety of chicken does Angus have?

9. Solve the equation 2(x + 4) - 3(x - 2) = 16.

10. This is a version of a puzzle written by Henry Dudley in 1907:

Look at the following arrangement of numbers:

7   28   196   34   5

Notice that 7 · 28 = 196, but that 34 · 5 does not equal 196.

It's possible to rearrange those numbers so that when each two-digit number is multiplied by its one-digit neighbor, the result is the three-digit number in the middle...

(I worked on this one for quite some time and I ultimately found that prime factorizing was very useful. It's possible that there is more than one correct answer.)

4 FRACTIONAL EXPONENTS

So, you've dealt with negative exponents, and a reasonable next question might be, "Can you have fractional exponents?" The answer is yes. Otherwise this would be a very short lesson. Indeed, the concept behind fractional exponents is mathematically very important, although, as you'll see shortly, mathematicians generally work with that concept without actually using fractional exponents. (I know that sounds a little crazy, but you'll see what I mean.)

So, what does, for example, $x^{1/2}$ mean?

1. **Actually, using the Laws of Exponents, you can answer that question yourself. Whatever $x^{1/2}$ means, use the First Law to figure out what the simplified version of this expression is:**

$$(x^{1/2})(x^{1/2})$$

In fact, the answer to Problem 1 is really the definition of $x^{1/2}$: $x^{1/2}$ is the thing that, when multiplied by itself, gives you **x**. In other words, when you square $x^{1/2}$, you get **x**, and $x^{1/2}$ is called *the square root of* **x**. Generally speaking, the square root of **x** is not written with a fractional exponent, but instead is written like this:

$$\sqrt{x}$$

The $\sqrt{}$ symbol is called a radical sign. When I was your age, the word *radical* was slang for "awesome," but the mathematical version gets its name from the Latin word *radix*, which means "root." (It's the same Latin word from which the English word *radish* comes.) Read aloud, $\sqrt{x}$ is "the square root of **x**," or often, for short, just "root **x**." Although you will see $\sqrt{x}$ far more often than you will see $x^{1/2}$, it's important that you know that $x^{1/2}$ and $\sqrt{x}$ are two ways of writing exactly the same concept.

Other fractions beside ½ are perfectly fair game in exponents. For instance, you can have $x^{1/3}$.

2. **Use the First Law to simplify the expression $(x^{1/3})(x^{1/3})(x^{1/3})$.**

So, $x^{1/3}$ is the thing that has to be multiplied by itself three times (that is, it has to be cubed) in order to give you **x**, and it's called *the cube root of* **x**. It's usually written like this:

$$\sqrt[3]{x}$$

By the same logic, $x^{1/4}$, $x^{1/5}$, $x^{1/6}$, and so on are all perfectly legitimate mathematical expressions. Those three would be read, "the fourth root of **x**, the fifth root of **x**, and the sixth root of **x**," and they'd usually be written $\sqrt[4]{x}$, $\sqrt[5]{x}$, and $\sqrt[6]{x}$.

If you wanted to get fancy, you could use a fractional exponent with a numerator other than one, such as $x^{2/4}$. So what might $x^{2/4}$ mean? Well, the **2** in the numerator of the exponent means "squared" and the **4** in denominator of the exponent means "take the fourth root." There are at least three ways to simplify $x^{2/4}$ into something you already know, and I'm going to ask you to try all three. You should get the same answer using each method, but I want you to see how and why they work.

3. **The first way to think about simplifying $x^{2/4}$ is to read it like so: "Take the fourth root of x and square it." That's the same as thinking of it this way: $\left(\sqrt[4]{x}\right)^2$. Remember that if you multiplied the fourth root of x by itself four times, you'd get x. So the question is, what do you get when you multiply the fourth root of x by itself only twice?**

4. **The second way to think about it is to read $x^{2/4}$ as, "Square x, then take the fourth root of that," or $\sqrt[4]{x^2}$. So, what could you multiply by itself four times and end up getting x^2?**

5. **The third method of simplifying $x^{2/4}$ is by far the easiest. Just simplify the fraction in the exponent.**

From now on you should feel free to go ahead and use the third method of just simplifying the fraction in order to simplify expressions like $x^{2/4}$. I just wanted you to try the other two ways of thinking about it so that you'd be sure that the third way really works.

Of course, if you have a number with a fractional exponent instead of a variable, it's often possible to simplify the expression quite a bit. For instance, $27^{2/3}$ is equal to **9**. I'll let you figure out how I knew that as you work on the next set of problems. (You could also think of it as a hint for how to do some of them.)

Simplify each of the following expressions. (Assume that all roots are positive — I'll say more about that in a moment.)

6. $16^{1/2}$

7. $81^{1/4}$

8. $49^{3/2}$ **(In this case you can either do the cubing or the square-rooting first. One method is quite a bit easier...)**

9. $64^{1/2}$

10. $64^{1/3}$

11. $64^{1/6}$

12. $4^{5/2}$

13. $8^{2/3}$

Expressions with fractional exponents also obey the Laws of Exponents perfectly, so...

Use the Laws of Exponents to simplify the following expressions:

14. $x^{2/3} \cdot x^{5/4}$

15. $\dfrac{x^{9/13}}{x^{7/13}}$

16. $(x^{1/2})^{3/2}$

17. $(x^{2/3} y^{1/2})^{1/3}$

Just to repeat what I said earlier, the truth is that in most math textbooks (including the rest of this one), you'll pretty rarely see fractional exponents. For the most part, you'll see radical signs instead — but you'll know that they mean the same thing as fractional exponents.

For the rest of this lesson (and actually the two lessons that follow), I'll be asking you to deal entirely with square roots: no more cube roots or fourth roots or so on.

So, here's the first important thing to know about square roots. Consider an expression such as $\sqrt{16}$. Actually, the number **16** has two square roots. **(4)(4) = 16**, so **4** is a square root of **16**. But **(-4)(-4) = 16** as well, so **-4** is also a square root of **16**. Well, in order not to have to deal with having two options all of the time (because *every* positive number has two square roots), mathematicians have agreed that the expression $\sqrt{16}$ will only refer to the *positive* square root of **16**. So $\sqrt{9}$ = **3**, not **-3**, and $\sqrt{x^2}$ = x, not **-x**. If you want to indicate the negative square root of **16**, you write it like this:

$-\sqrt{16}$

This will make your life a lot simpler.

The second thing I'd like you to realize about square roots is that most square roots are not neat and tidy numbers. The square root of **16** is exactly **4**. The square root of **9** is exactly **3**. The square root of **4** is exactly **2**. But most square roots are not as well behaved as that. Think about, for instance, $\sqrt{2}$. The square root of **2**, unlike the square roots of **16** or **9** or **4**, is not a whole number. In fact, the square root of **2** is even more ill behaved than simply not being a whole number.

Back in *Jousting Armadillos*, you had a short lesson called "Ratios" that introduced you to the idea of rational and irrational numbers. Remember that rational numbers were ones that could be expressed as ratios: they can be written as fractions. So, **7** is a rational number (it can be written **7/1**), and **3/5** is obviously a rational number, and so is **2.25**. (It's equal to **2 ¼** or **9/4**.) Irrational numbers, on the other hand, are ones like π that can't be written as fractions.

Well, $\sqrt{2}$ is an irrational number like π, and so are most square roots. Now, irrational numbers are beautiful and useful things (though possibly dangerous: at least one ancient Greek was supposedly killed over them), but they do have one difficult quality, which is that they can't be written exactly as fractions or decimals because their decimal versions go on forever without repeating. That's why the symbol π is used.

In the case of square roots, what this means is that when you see $\sqrt{16}$ or $\sqrt{25}$, for instance, you should get used to automatically simplifying them to **4** or **5**. The numbers **16** and **25** are called *perfect squares* (or sometimes just *square numbers*) precisely because they have whole-number square roots.

But when you see $\sqrt{2}$, you can't simplify it any further than that. The same thing is true for $\sqrt{7}$, $\sqrt{13}$, and many other square roots. (In the next lesson, we'll take a look at how, for example, $\sqrt{8}$ can be simplified somewhat, even though **8** is not a perfect square.)

The third thing you ought to know right now about square roots is whether or not this is an okay thing to write:

$$\sqrt{-1}$$

18. **Does it seem to you that -1 can have a square root? Why or why not?**

I imagine that you may have concluded that $\sqrt{-1}$ doesn't make a whole lot of sense. You would have very good reasons for thinking that. After all, you know quite well that any number, positive or negative, when multiplied by itself gives you a positive number. Well, as it turns out, $\sqrt{-1}$ actually is a perfectly legitimate mathematical concept, but you may be relieved to hear that you need to know very little about it for the time being.

There are just a couple of things I would like you to know for right now having to do with $\sqrt{-1}$. As you already know, numbers fall into different categories. I've just been talking about whole numbers, rational numbers, and so on. Some of those categories overlap. For instance, whole numbers are the group **0, 1, 2, 3, 4, 5**, and so on, and integers are all of those plus the equivalent negatives: **-1, -2, -3, -4, -5**, and so on. Therefore, all whole numbers are integers, but not all integers are whole numbers.

If you were to make a chart of the varieties of numbers that you've encountered so far, it might look like what I've drawn below. (I've given examples of numbers that fall into each category, and if one category includes the other I've put the boxes inside each other.)

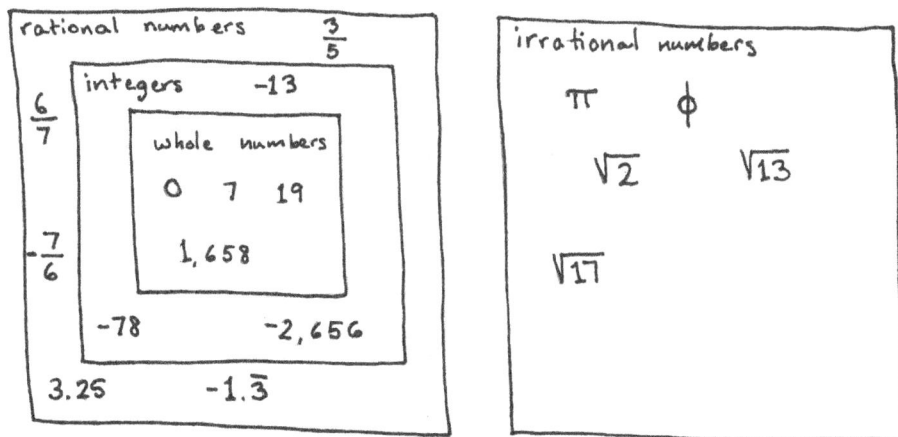

rational numbers $\quad \dfrac{3}{5}$

integers $\quad$ −13

whole numbers

0 $\quad$ 7 $\quad$ 19

1,658

−78 $\qquad$ −2,656

$\dfrac{6}{7}$

$-\dfrac{7}{6}$

3.25 $\qquad$ −1.$\overline{3}$

irrational numbers

$\pi \qquad \phi$

$\sqrt{2} \qquad \sqrt{13}$

$\sqrt{17}$

Notice that rational numbers and irrational numbers are two separate categories that don't overlap. Well, at some point mathematicians decided that they would call the *entire group* of numbers, including rational and irrational numbers, the "real numbers." So the picture looks like this:

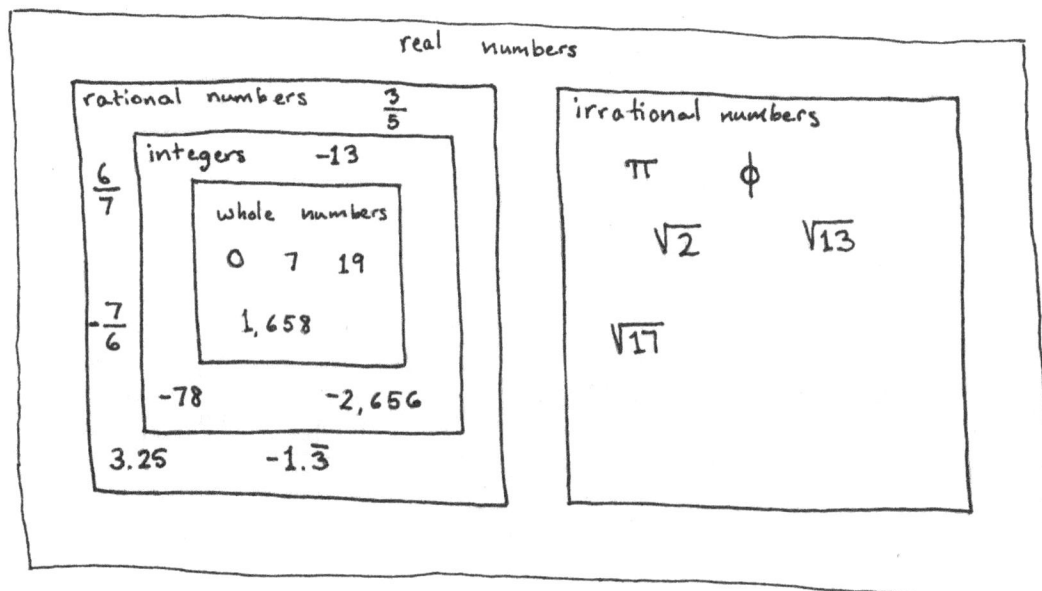

But in the 1500s, mathematicians began to experiment with a whole new group of numbers that didn't fit into any of those categories: the square roots of negative numbers. The square roots of negative numbers turned out to be very useful and totally fascinating. In 1637, René Descartes (you remember him, right? from the Cartesian coordinate plane in *Crocodiles & Coconuts*?) gave this new category of numbers a name...

19. **Don't look at the next page. Based on the chart above, what do you suppose Descartes decided to call those numbers?**

Well, if those numbers aren't "real," obviously they must be... *imaginary*. This was either the coolest bit of mathematical naming ever or the worst. I think it's cool because it's wonderfully poetic, but on the other hand, calling these numbers "imaginary" tends to give the impression that they aren't, well, real. And while it's true that they don't fall into the mathematical category of "real numbers," they are very real in the ordinary sense of actually existing. In fact, as I say, they are pretty wonderful.

However, you won't really study imaginary numbers for quite a while yet. So, other than reminding you of the categories of numbers (which you ought to know), the whole section about imaginary numbers that you just read really means this for right now: in this book, whenever you see a variable under a radical sign, such as $\sqrt{x}$, you get to assume the variable can represent only positive numbers.

In this lesson, there's just one thing I'd like you to learn to actually *do* with square roots, and that's to estimate the decimal value of non-perfect square roots. (Remember, they don't have exact decimal values.) By the way, I want to be sure that you know that when you use a calculator to find the value of, say, $\sqrt{2}$, what the calculator has actually done is estimate the value of $\sqrt{2}$; the actual value of $\sqrt{2}$ cannot be written as a decimal.

20. In order to be able to estimate the values of square roots without using a calculator, you need to know the numbers that are perfect squares, and what their square roots are. So, copy this chart into your notebook and finish filling it out:

perfect square	1	4	9										
square root	1	2	3	4	5	6	7	8	9	10	11	12	13

21. Any competent mathematician should have the perfect squares and their square roots memorized at least through the number whose square root is 13. So memorize that chart. If you want to memorize higher than that, it's up to you, but you need to have at least that much available in your head at any time.

22. Now you'll use that memorized chart to estimate square roots. For instance, let's consider $\sqrt{20}$. Now, 20 is not a perfect square, so $\sqrt{20}$ is an irrational number — it has no exact fractional or decimal equivalent. But, what two perfect squares does 20 lie between? Therefore, what two whole numbers must $\sqrt{20}$ lie between?

That's all there is to it. In the case of $\sqrt{20}$, 20 is pretty much halfway between **16** and **25**, so $\sqrt{20}$ is probably close to halfway between **4** and **5**. (In fact, the first six digits of the decimal equivalent of $\sqrt{20}$ are **4.47213**). And all you really need to be able to say is something like, "$\sqrt{20}$ is between **4** and **5**, probably close to halfway." In the case of $\sqrt{24}$, 24 is still between **16** and **25**, but it's much closer to **25**, so you ought to be able to say something like, "$\sqrt{24}$ is between **4** and **5**, closer to **5** than to **4**."

Make estimates for the following square roots:

23. $\sqrt{7}$ 24. $\sqrt{13}$

25. $\sqrt{17}$ 26. $\sqrt{37}$

27. $\sqrt{45}$ 28. $\sqrt{73}$

29. $\sqrt{113}$ 30. $\sqrt{150}$

31. $\sqrt{170}$

32. Write a *Note to Self* about *roots*. It ought to explain what roots are; make sure to explain the relationship between roots and fractional exponents and give a couple of examples of estimating the decimal value of square roots.

33. The state police (and other agencies), when they arrive on the scene of a high-speed traffic accident, can use a formula to determine the speed of the vehicles involved in the accident. They do this by measuring the length of the mark left by the skidding car's tires. The formula they use looks like this: $S = \sqrt{20D}$. The S represents the car's speed in miles per hour and the D represents the distance or length of the skid mark. Use this equation to determine the following:

a) How fast was a car going if it left a mark of 5 feet? 20? 45? 125?
b) This formula implies that the only factor that determines the length of the skid marks is the speed of the car. What are some other factors that seem as if they might have an influence on the length of the skid marks?

REVIEW

1. Graph the equation $4x^2 + 9y^2 = 36$.

2. Make an approximate copy of the graph below in your notebook and add two points to it: one using interpolation and one using extrapolation. Explain which point is which.

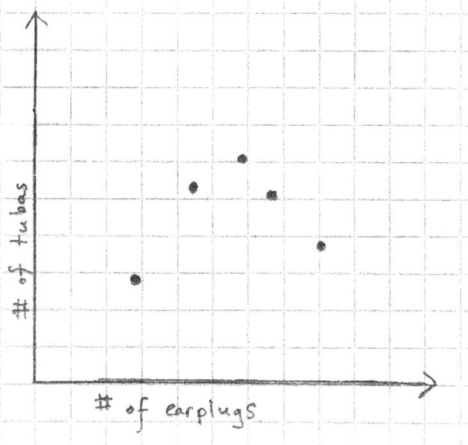

3. Solve the equation $\frac{3x - 3}{-x + 1} = -3$ and explain what the solution means.

4. Thomas and Leo spent a week collecting rubber ducks. On the first day, they acquired one-twelfth of the total number that they would ultimately collect. On Day 2 they acquired a further third of their final total. On Day 3 they acquired one-sixth of the grand total. On Day 4, they collected 30 ducks. On Day 5 they got hold of another quarter of the final total. On Day 6, Leo mysteriously lost six ducks. On the last day, they acquired one-eighth of the final total. How many rubber ducks did they collect in all?

5. Do the following division problem, rounding the answer to the nearest hundredth:

$9{\overline{\smash{)}25}}$

6. Find the greatest common factor of $84x^3$ and $90xy$.

7. Simplify the following expression:

$$\frac{\dfrac{15x^3}{2y}}{\dfrac{3x}{4y^2}}$$

8. Express 64 as a power of 2.

9. Simplify the following expression:

$$\frac{3|2-7|+15\cdot2}{-11-(-3\cdot2)}$$

10. Here is a classic puzzle:

You need exactly four gallons of water from the well. (Don't ask me why — you just need exactly four, okay?) You own one jug that holds exactly five gallons and one jug that holds exactly three gallons. Neither jug has any markings on it. How are you going to get those four gallons? (By the way, the well contains plenty of water, so it's okay if you have to waste some in the process. But try to at least pour it on some thirsty plants, okay?)

5 MANIPULATING SQUARE ROOTS

The first thing to realize about square roots, as I mentioned in the last lesson, is that they obey all of the Laws of Exponents — they're the same thing as fractional exponents, after all. Rewritten for square roots, the First Law would look like this:

$$(\sqrt{x})(\sqrt{x}) = x$$

Notice that the left-hand side of that equation means "**root x** times **root x**." To indicate multiplying square roots you can also just put them next to each other, like this:

$$\sqrt{x}\sqrt{x} = x$$

The Second and Third Laws are not going to be especially useful for the stuff that you'll do in this lesson, but the Fourth and Fifth Laws definitely will be.

1.   **The Fourth Law of Exponents tells us that $(xy)^a = x^a y^a$. According to the Fourth Law, what must $(xy)^{1/2}$ be equal to?**

2.   **$(xy)^{1/2}$ is the same thing as $\sqrt{xy}$, so what is $\sqrt{xy}$ equal to?**

The Fourth Law can be quite helpful in simplifying expressions that involve square roots, especially when you realize that you can apply it in either direction — that is, you can use it to change the expression $\sqrt{2}\sqrt{8}$ into $\sqrt{16}$ or you can use it to change the expression $\sqrt{9x}$ into $\sqrt{9}\sqrt{x}$.

3.   **Can you simplify $\sqrt{2}$?**

4.   **At this point, can you simplify $\sqrt{8}$?**

5.   **You might think that you therefore can't simplify $\sqrt{2} \cdot \sqrt{8}$. But apply the Fourth Law to it just as I did in the last paragraph, and you can. Go ahead and do so.**

Use what you just learned to simplify the following expressions as far as possible:

6. $\sqrt{20} \cdot \sqrt{5}$

7. $\sqrt{12} \cdot \sqrt{3}$

8. $\sqrt{4} \cdot \sqrt{2} \cdot \sqrt{18}$ **(There are at least two ways to do this one, both of which lead to the same result.)**

You can use the Fourth Law in the other direction (if you see what I mean) to simplify an expression like $\sqrt{8}$. Doing so requires recognizing that **8** has a *factor* that is a perfect square. In other words, $\sqrt{8}$ can be rewritten as $\sqrt{4 \cdot 2}$.

9. Apply the Fourth Law to rewrite the expression $\sqrt{4 \cdot 2}$ as the product of two square roots. One of those two square roots can be simplified, so simplify it. You should be left with an expression that looks like a whole number multiplied by a number under a square root. That's as simple as the expression can get.

Rewriting an expression such as $\sqrt{8}$ in the way you just did is called writing it in its *simplest radical form*. It is probably not immediately obvious to you why $2\sqrt{2}$ is simpler than $\sqrt{8}$. It's pretty easy to estimate the size of either one of those expressions. $\sqrt{8}$ is somewhere between **3** and **4** (because, as you'll recall from the last lesson, 8 is between **4** and **9**). $\sqrt{2}$ is somewhere between **1** and **2**, so $2\sqrt{2}$ is probably somewhere between (surprise!) **3** and **4**. However, writing expressions in their simplest radical form can be useful. For instance, it can help you get a sense of the size of an expression like $\sqrt{432}$, which is pretty hard to do otherwise unless you memorized the perfect squares beyond **169**.

10. Rewrite $\sqrt{432}$ by factoring 432. (You don't have to factor 432 completely. One of the square numbers that you memorized in the last lesson is a factor of 432, which is the point. Once you figure out that factor, use it to rewrite $\sqrt{432}$.) Now rewrite that expression using the Fourth Law and simplify the part that can be simplified so that you end up with an answer in the same sort of form as $2\sqrt{2}$.

It's pretty hard to estimate the value of $\sqrt{432}$. But estimating the value of $12\sqrt{3}$ is not all that hard. $\sqrt{3}$ is between **1** and **2**, closer to **2** than **1**, so $12\sqrt{3}$ is probably a little less than **24**.

Rewrite each of the following expressions in simplest radical form.

11. $\sqrt{40}$ 12. $\sqrt{150}$

13. $\sqrt{72}$ 14. $\sqrt{810}$

15. $\sqrt{324}$ 16. $\sqrt{95}$

As I often do, I threw in a couple of slightly different ones at the end. In Problem 15, it probably wasn't apparent right away that **324** is a perfect square, but when you factored it into **81 · 4**, I hope you saw that it turned out to be one. And, of course, Problem 16 was there just to remind you that not every square root can be simplified using this method — only ones that have a factor that's a perfect square. $\sqrt{95}$ is already in simplest radical form. Sometimes I can be a little bit of a pain, but you know that by now.

It's also sometimes possible to simplify radical expressions with variables in them. (Keep in mind that whenever you see a variable under a radical sign — at least in this book — you can assume that the variable is positive.) Simplifying an expression like, for instance, $\sqrt{x^6}$ is pretty simple. All you really have to do is ask yourself, "What times itself would give me x^6?"

17. What times itself would give you x^6?

18. What's the simplified version of $\sqrt{x^4}$?

19. What's the simplified version of $\sqrt{x^8}$?

20. What's the simplified version of $\sqrt{x^{10}}$?

21. How do you simplify the square root of any variable raised to an even power?

Based on what you discovered, you might be tempted to conclude that you can't simplify the square root of a variable raised to an odd power, such as $\sqrt{x^3}$. However, it is possible to simplify $\sqrt{x^3}$ in a way that is quite similar to what you just did in Problems 11 – 16.

22. x^3 can be factored. What are its factor pairs?

23. Remember that when you simplified, say, $\sqrt{8}$ into simplest radical form, you did so by factoring 8 into 4 · 2, and you were able to simplify that because 4 is a perfect square. Well, one of the factors of x^3 is also a perfect square, so you can use the same technique that you used to simplify $\sqrt{8}$. Go ahead and do so — just as you did when you simplified $\sqrt{8}$ to $2\sqrt{2}$, you should end up with something still under the radical sign.

That same technique allows you to simplify the square root of any variable raised to an odd power.

Rewrite the following expressions in simplest radical form:

24. $\sqrt{x^5}$

25. $\sqrt{x^{11}}$

26. $\sqrt{x^{27}}$

You can use what you now know to simplify all sorts of square roots that contain combinations of numbers and variables. For instance, take the expression $\sqrt{50x^5}$. In order to put it in simplest radical form, you are looking for factors that are perfect squares. In this case, you've got **25** and $\mathbf{x^4}$, so you'd rewrite $\sqrt{50x^5}$ as $\sqrt{25 \cdot 2 \cdot x \cdot x^4}$. The Fourth Law then allows you to rewrite $\sqrt{25 \cdot 2 \cdot x \cdot x^4}$ as $\sqrt{25}\sqrt{x^4}\sqrt{2x}$. And $\sqrt{25}\sqrt{x^4}\sqrt{2x}$ can be further simplified to $5x^2\sqrt{2x}$, which is as simple as that expression can get.

Rewrite the following expressions in simplest radical form:

27. $\sqrt{81y^2}$ 28. $\sqrt{200x^4}$

29. $\sqrt{80x^3}$ 30. $\sqrt{50x^5}$

31. $\sqrt{8x^4y^3}$ 32. $\sqrt{32x^3y^2}$

33. $\sqrt{96x^2y^6z^7}$

When you encounter an expression with square roots that you have to simplify, you need to think flexibly and be willing to apply the Fourth Law in both directions. Sometimes it will make sense to do some multiplying and combining of expressions under square roots in order to simplify them. Take, for example, the expression $4\sqrt{x^3} \cdot 3\sqrt{x^7}$. Now, you could choose to simplify $4\sqrt{x^3}$ and $3\sqrt{x^7}$ separately and multiply the resulting expressions, and in fact that route will get you to the simplest radical form. But in this case it's easier to multiply and apply the Fourth Law first, thereby changing $4\sqrt{x^3} \cdot 3\sqrt{x^7}$ into $12\sqrt{x^{10}}$. (The Fourth Law tells us that $\sqrt{x^3} \cdot \sqrt{x^7} = \sqrt{x^{10}}$.) And $12\sqrt{x^{10}}$ is easy to simplify to **$12x^5$**.

On the other hand, sometimes it will make more sense to simplify the things under the radical sign first, then multiply and, if necessary, keep simplifying. Take, for example, the expression $\sqrt{48x^3} \cdot \sqrt{72x}$. One route would be to multiply the two expressions to get $\sqrt{3{,}456x^4}$ (**48** times **72** is **3,456**). But finding factors of **3,456** that are perfect squares would be a royal pain. In this case, you'd be way better off finding the perfect square factors first, like so:

$$\sqrt{48x^3} \cdot \sqrt{72x} = \sqrt{16 \cdot 3x^3} \cdot \sqrt{36 \cdot 2x}$$

... and then simplifying each of those separate radicals, like this:

$$\sqrt{16 \cdot 3x^3} \cdot \sqrt{36 \cdot 2x} = 4\sqrt{3x^3} \cdot 6\sqrt{2x}$$

... then multiplying:

$$4\sqrt{3x^3} \cdot 6\sqrt{2x} = 24\sqrt{6x^4}$$

... and simplifying once more to the simplest form: $24\sqrt{6x^4} = 24x^2\sqrt{6}$.

Like I say, you've got to think flexibly and try different tactics.

Convert the following expressions to simplest radical form. In several cases there are multiple correct paths that you could take to simplest radical form.

34. $4\sqrt{x} \cdot 2\sqrt{x^5}$

35. $\sqrt{x} \cdot \sqrt{x} \cdot \sqrt{x}$

36. $\sqrt{32x} \cdot \sqrt{18x^4}$

37. $\sqrt{5x^2} \cdot \sqrt{5}$

38. $\sqrt{121x^5} \cdot \sqrt{360x^3}$

39. $2\sqrt{3x^1} \cdot 3\sqrt{3x^3}$

There is one more subject that we need to deal with in this lesson, which has to do with expressions that have fractions under radical signs. The basic rule for converting fractions under radical signs into simplest radical form is that you want to end up with an expression that doesn't have a radical sign in the denominator.

In order to see why this would be the case, let's think back to estimating the values of square roots. Suppose we were dealing with the expression $\sqrt{\dfrac{10}{3}}$. Estimating the value of that expression seems very hard to me. In fact, at first I might draw a little bit of a blank. After a moment, I might realize that $\sqrt{\dfrac{10}{3}} = \dfrac{\sqrt{10}}{\sqrt{3}}$. (This is an application of the Fifth Law of Exponents, which tells me that $\left(\dfrac{x}{y}\right)^a = \dfrac{x^a}{y^a}$ as long as **y** is not equal to zero). But it's still pretty hard to estimate the value of $\dfrac{\sqrt{10}}{\sqrt{3}}$. $\sqrt{10}$ is between **3** and **4** and $\sqrt{3}$

is between **1** and **2**, but it kind of makes my brain hurt to try to estimate what I'll get when I divide something that I don't know the exact value of by something else that I don't know the exact value of.

Suppose, on the other hand, that I wanted to estimate the value of $\dfrac{\sqrt{30}}{3}$. I think this is easier: $\sqrt{30}$ is between **5** and **6**, so if I'm dividing it by **3**, I'll get something that's a bit less than **2**.

Well, as you'll shortly discover, $\sqrt{\dfrac{10}{3}}$ is equal to $\dfrac{\sqrt{30}}{3}$.

40. **As I already pointed out, $\sqrt{\dfrac{10}{3}}$ is equal to $\dfrac{\sqrt{10}}{\sqrt{3}}$. Now, you know that you can multiply any fraction by 1 and get an equivalent fraction. (For example, just to refresh your memory, $\left(\dfrac{2}{3}\right)\left(\dfrac{5}{5}\right) = \dfrac{10}{15}$. Because $\dfrac{5}{5}$ is just another way writing 1, $\dfrac{2}{3}$ and $\dfrac{10}{15}$ are two ways of writing the same fraction.) In the case of $\dfrac{\sqrt{10}}{\sqrt{3}}$, the version of 1 that you want to multiply it by is $\dfrac{\sqrt{3}}{\sqrt{3}}$. Go ahead and do so.**

I hope you can see that the fraction you ended up with in Problem 40, $\dfrac{\sqrt{30}}{\sqrt{9}}$, has a perfect square under the radical sign in the denominator (that was the whole point of multiplying $\dfrac{\sqrt{10}}{\sqrt{3}}$ by $\dfrac{\sqrt{3}}{\sqrt{3}}$) and that it's equal to $\dfrac{\sqrt{30}}{3}$.

I think this is actually a pretty simple and sensible technique, but here's a little word of advice: before using it, always simplify the original fraction if you can.

Convert the following expressions to simplest radical form.

41. $\sqrt{\dfrac{5}{2}}$

42. $\sqrt{\dfrac{11}{6}}$

43. $\sqrt{\dfrac{20}{12}}$

44. $\sqrt{\dfrac{10}{8}}$

45. $\sqrt{\dfrac{16}{3}}$

Once again, I was changing things up a little bit in those last two problems. Let's look at Problem 45 first. If you got $\dfrac{\sqrt{48}}{3}$ as your answer, you were correct, but not as correct as you could have been — in the sense that this answer can be simplified further. The way you would have arrived at $\dfrac{\sqrt{48}}{3}$ was by multiplying $\sqrt{\dfrac{16}{3}}$ by $\dfrac{\sqrt{3}}{\sqrt{3}}$ exactly as I showed you how to do. However, the numerator of $\sqrt{\frac{16}{3}}$ is a perfect square, right? (And some of you may have spotted this.) So, you can simplify $\sqrt{\frac{16}{3}}$ directly to $\frac{4}{\sqrt{3}}$, and when you multiply that by $\frac{\sqrt{3}}{\sqrt{3}}$ in order to get rid of the radical sign in the denominator, you get $\frac{4\sqrt{3}}{3}$ — which is the simplest (and therefore, in some sense, the most correct) version.

46. **However, it doesn't really matter if you got $\frac{\sqrt{48}}{3}$ because it's pretty simple to change that into $\frac{4\sqrt{3}}{3}$. Explain why $\frac{\sqrt{48}}{3}$ and $\frac{4\sqrt{3}}{3}$ are equal.**

Okay, now let's look at Problem 44. In that case, you may have gotten $\frac{\sqrt{80}}{8}$, which is correct, but is not the simplest version. If you remembered to simplify the fraction first, you may also have gotten $\frac{\sqrt{20}}{4}$, which is also correct, but still not the simplest version. Once you simplified $\sqrt{\frac{10}{8}}$ to $\sqrt{\frac{5}{4}}$ (assuming you remembered to do that), you may have recognized that **4** is a perfect square and therefore you may have gone directly to the simplest version, which doesn't even require multiplying by anything: $\frac{\sqrt{5}}{2}$.

47. **Explain why $\frac{\sqrt{80}}{8}$ and $\frac{\sqrt{20}}{4}$ are both equal to $\frac{\sqrt{5}}{2}$. (This is slightly trickier than Problem 59, because in this case you have to put the numerators in simplest radical form and then recognize that the resulting fractions can be further simplified.)**

If you remember to simplify fractions like $\sqrt{\frac{10}{8}}$ before you try to get rid of the radical sign in the denominator and you recognize when you have perfect squares, you shouldn't have to deal with simplifying expressions like $\frac{\sqrt{80}}{8}$ or $\frac{\sqrt{20}}{4}$, but just in case you forget, here are a few to practice on…

Convert the following expressions to simplest radical form. You'll have to convert the numerators to simplest radical form and then decide whether or not the fractions can be simplified further.

48. $\dfrac{\sqrt{12}}{6}$

49. $\dfrac{\sqrt{63}}{9}$

50. $\dfrac{\sqrt{45}}{7}$

51. $\dfrac{\sqrt{125}}{5}$

The last thing I'll ask you to do in this lesson is to simplify a few radical expressions that involve fractions and variables. The goal is still the same (no radicals in the denominator) and the basic technique is still the same (multiply the fraction by something so that you get a perfect square in the denominator).

For instance, if you were trying to express $\dfrac{\sqrt{5}}{\sqrt{x}}$ in simplest radical form, the question is:

"What fraction equal to **1** do I need to multiply $\dfrac{\sqrt{5}}{\sqrt{x}}$ by in order to get a perfect square under the radical sign in the denominator?"

Answer that question and the rest is easy.

Rewrite the following expressions in simplest radical form. It's still advisable to simplify the fractions first when possible. At any rate, make sure that your final version has the numerator in simplest radical form and you've simplified the fraction if possible.

52. $\dfrac{\sqrt{6}}{\sqrt{x}}$

53. $\dfrac{\sqrt{10}}{\sqrt{8x}}$

54. $\dfrac{\sqrt{9}}{\sqrt{11x}}$

55. $\dfrac{\sqrt{16}}{\sqrt{3xy}}$

56. $\dfrac{\sqrt{14}}{\sqrt{7x^2y}}$

57. It's time for a **Note to Self** on the subject of **simplest radical form.** If I were writing this Note, I think I would choose several examples and clearly explain how to translate them into simplest radical form: one example like Problems 11 through 20, one like Problems 31 through 42 (I'd pick one where a variable is raised to an odd power, since those are slightly trickier), one like Problems 43 through 50, and one like Problems 52 through 73 (or possibly two different ones — one with a variable and one without).

REVIEW

1. Find the least common multiple of 225 and 105.

2. Simplify the following expression:

$$\frac{x}{3y} - \frac{2x}{5y}$$

3. Solve the following inequality:

$$\frac{-6x + 2}{5} \geq 4$$

4. Graph the following equation:

$$xy = 6$$

Solve the following pairs of simultaneous equations:

5. $6x + y = 4$
$-10x + 3y = 40$

6. $2y = 3x - 5$
$-6x + 4y = -10$

7. $6x - 5y = -8$
$3x + y = 3$

8. The value of Linus's model moose collection plummeted by 23% in a single week. (Boy, that turned out to be a lousy investment.) If his collection was worth $269.50 at the end of the week, what had it been worth at the beginning?

9. Three years ago, Gregor Samsa was one-third his brother's age. In four years, he'll be half his brother's age. How old are the two brothers now?

10. This is another classic puzzle:

You have one barrel of wine and one barrel of water. Each barrel contains exactly the same volume of its respective liquid. You fill a glass with water from the water barrel and empty the glass into the wine barrel. Then you fill the same glass with the wine-and-water mixture and pour it back into the water barrel. Now, is there more wine in the water barrel or more water in the wine barrel?

6 SOLVING RADICAL EQUATIONS

Now that you can simplify expressions that involve square roots, we'll take a look at how to solve equations that involve square roots. But before we do that, I want to talk about how square roots are involved in addition and subtraction. Now, you know that in the case of multiplication, this is true: $\sqrt{x}\sqrt{y} = \sqrt{xy}$

And you know that in the case of division, this is true: $\dfrac{\sqrt{x}}{\sqrt{y}} = \sqrt{\dfrac{x}{y}}$

Because those two things are true, you may be quite reasonably tempted to believe that something similar would be true about addition and subtraction. So I want to be very clear about the following:

It is NOT true that $\sqrt{x+y} = \sqrt{x} + \sqrt{y}$ **or that** $\sqrt{x-y} = \sqrt{x} - \sqrt{y}$.

I think you can prove to yourself pretty quickly that neither of those statements is true. (Or at least they're not true for all possible values of **x** and **y** the way that $\sqrt{x}\sqrt{y} = \sqrt{xy}$ and $\dfrac{\sqrt{x}}{\sqrt{y}} = \sqrt{\dfrac{x}{y}}$ are.)

1. **Simplify the expression** $\sqrt{9} + \sqrt{16}$ **by taking the square roots of 9 and 16 and then adding them.**

2. **Simplify the expression** $\sqrt{9+16}$ **by adding 9 and 16 and then taking the square root.**

3. **Is the answer to Problem 2 the same as the answer to Problem 1?**

I hope you'll agree that this example proves that $\sqrt{x} + \sqrt{y}$ is not equal to $\sqrt{x+y}$.

4. **Simplify the expressions** $\sqrt{25} - \sqrt{16}$ **and** $\sqrt{25-16}$ **as you did in Problems 1 and 2 in order to check whether** $\sqrt{25} - \sqrt{16}$ **is equal to** $\sqrt{25-16}$.

And I hope you'll agree that that example proves that $\sqrt{x} - \sqrt{y}$ is not equal to $\sqrt{x-y}$.

5. **I've seen students make mistakes around this issue often enough that I'm going to ask you to devote a page in your Note to Self book simply to copying down this statement:**

It is NOT true that $\sqrt{x+y} = \sqrt{x} + \sqrt{y}$ **or that** $\sqrt{x-y} = \sqrt{x} - \sqrt{y}$.

Underline it several times or put stars around it or something.

The fact that you just wrote down in your Note to Self book means that none of the following expressions can be simplified:

$\sqrt{5} + \sqrt{7}$

$3\sqrt{2} - 4\sqrt{3}$

$\sqrt{n} + 5\sqrt{m}$

$7\sqrt{x} - 3\sqrt{y}$

In each of those cases, the expressions are already as simple as you can get. On the other hand, it's not true that you can *never* simplify an expression that involves radicals and addition or subtraction. For example, here are some similar expressions, all of which can be simplified further:

$$\sqrt{5} + \sqrt{5}$$
$$3\sqrt{3} - 4\sqrt{3}$$
$$\sqrt{n} + 5\sqrt{n}$$
$$7\sqrt{y} - 3\sqrt{y}$$

6. **Examine the list of expressions that can't be simplified further and the list that can be simplified further. What is the key difference between them?**

The reason that you can't simplify $\sqrt{n} + 5\sqrt{m}$ any further is simply that n and m are not the same and therefore $\sqrt{n}$ and $\sqrt{m}$ are not the same — it's like adding apples and wingnuts, as my co-author Greg likes to say. (Since n and m are variables, yes, it is *possible* that they have the same value, but in general you should assume that they don't.) On the other hand, if you're adding $\sqrt{n}$ and $5\sqrt{n}$, you have no idea how big $\sqrt{n}$ is, but you know that you're adding one of those mystery-size things to five of them, and you must therefore have six of them: $\sqrt{n} + 5\sqrt{n} = 6\sqrt{n}$.

The same logic applies to subtracting $4\sqrt{3}$ from $3\sqrt{2}$. $\sqrt{3}$ and $\sqrt{2}$ are irrational numbers of different sizes. You can't subtract wingnuts from apples. But if you have three of something and you take away four of the same kind of thing, you're left with negative one of them: $3\sqrt{3} - 4\sqrt{3} = -\sqrt{3}$.

The one catch is that it may not be apparent whether an expression with addition or subtraction can be simplified or not until you've expressed all the parts of it in simplest radical form. For instance, you might not think at first glance that you'd be able to do anything with the expression $\sqrt{20} + \sqrt{5}$.

7. **Express $\sqrt{20}$ in simplest radical form. Once you do so, you'll see that you actually can simplify the expression $\sqrt{20} + \sqrt{5}$.**

Simplify the following expressions as far as possible, remembering to use simplest radical form:

8. $5\sqrt{x} + 5\sqrt{x}$ 9. $\sqrt{75} - 2\sqrt{3}$

10. $\sqrt{72} + \sqrt{18}$ 11. $\sqrt{28} + \sqrt{98}$

12. $\sqrt{18x} - \sqrt{50x}$ 13. $\sqrt{4x^3y} + \sqrt{9x^3y}$

That last one was a little tricky. I hope, once you got to $2x\sqrt{xy} + 3x\sqrt{xy}$, you were able to see that you were still adding two mystery things to three of the same kind of mystery thing — in this case the mystery thing in question was $x\sqrt{xy}$.

Sometimes it's useful to use the Distributive Property with expressions that involve square roots, and it's not especially tricky to do so as long as you remember what happens when you multiply square roots by other square roots. For example, consider $(\sqrt{x})(6 + \sqrt{x})$. Applying the Distributive Rule, you get $6\sqrt{x} + \sqrt{x^2}$, and $\sqrt{x^2}$ can be simplified to just x, so $(\sqrt{x})(6 + \sqrt{x})$ is equal to $6\sqrt{x} + x$. As with any case involving the Distributive Rule, one version is not necessarily preferable to the other, but you should know how to apply the rule.

You can also do what I've called "undistributing" (which I'll start calling by its proper name in the third chapter of this book). That is, if you see an expression like $16 + 4\sqrt{x}$, you should recognize that you could rewrite it as $4(4 + \sqrt{x})$.

Apply the Distributive Rule to the following expressions and simplify where appropriate:

14. $3(5x + \sqrt{x})$ 15. $\sqrt{7}(\sqrt{7} + 4)$

16. $2\sqrt{x} - 14$ 17. $\sqrt{x}(\sqrt{8x} + \sqrt{x})$

18. $3x\sqrt{5} + 3x\sqrt{7}$

Problems 19 – 27 are a mixed bag of the square root concepts covered in the last couple of lessons. If possible, simplify the following expressions:

19. $\sqrt{50}$ 20. $\sqrt{4x^4}$

21. $\dfrac{\sqrt{15}}{\sqrt{3}}$ 22. $5\sqrt{3} + 2\sqrt{3}$

23. $\sqrt{20x} + \sqrt{5x}$ 24. $(3\sqrt{2})^2$

25. $\sqrt{a^2 + b^2}$ 26. $(\sqrt{2})(\sqrt{18})$

27. $\sqrt{17} + \sqrt{7}$

Before we move on, I want to say a quick word about Problem 25. This kind of expression often fools people: they see things that are squared underneath a radical sign and they quite naturally want to simplify them. However, while $\sqrt{a^2 b^2}$ can definitely be simplified, $\sqrt{a^2 + b^2}$ is really an example of a $\sqrt{x + y}$ situation and it can't be simplified.

So, everything you've done in these last two lessons has really been just simplifying expressions. It's time now to look at solving single-variable equations that involve square roots — or *radical equations*, as they're called.

Here's an example of one:

$5\sqrt{x} = 35$

As you know very well, solving single-variable equations works according to one simple rule (whatever you do to one side of the equation, you need to do the same thing to the other side) and has one simple goal (to figure out what value or values of the variable make the equation true). This rule and this goal still apply with radical equations. So, the first step to solving the equation $5\sqrt{x} = 35$ is to divide both sides by **5**. Your work would look something like this:

$5\sqrt{x} = 35$
$\sqrt{x} = 7$

Now, there's just one more step to solving this equation. And that's to square both sides of the equation.

28. **But wait. That seems a little odd. If you're squaring both sides of the equation, you're multiplying each side by itself. Aren't you therefore multiplying each side by something different, and doesn't that break the one simple rule of solving single-variable equations? Explain why squaring both sides of an equation actually *does* follow that simple rule (which it does!).**

So, as I hope you just managed to prove to yourself, it's fine to square each side of the equation. Your work should look something like this:

$5\sqrt{x} = 35$
$\sqrt{x} = 7$
$x = 49$

In other words, solving radical equations works just the same as solving any other single-variable equation… with one important difference. I know that you always check to see whether the solution you've come up with in solving a single-variable equation makes the original equation true. (Honestly, this is almost always simple enough to do in your head and it's really worth doing. What's the square root of **49**? It's **7**. Does **5** times **7** equal **35**? Yup. That's all you need to do.) If the solution you came up with doesn't make the original equation true, then you made a mistake in your steps. *Except in the case of a radical equation: with a radical equation, you can do all of the steps perfectly and still end up with a solution that doesn't make the original equation true.* I'll show you.

29. **Solve the equation $\sqrt{x} + 10 = 6$. Now check the solution in the original equation. Double-check the steps that you used to solve the equation.**

You didn't do anything incorrectly; it's just that the equation $\sqrt{x} + 10 = 6$ doesn't have a solution. Or, more properly, I should say that it doesn't have a *real* solution. It does have an imaginary one. But, as I told you earlier, you don't need to worry about imaginary numbers yet.

Solve the following equations. Check your answers. Indicate if there is no real solution to an equation.

30. $3\sqrt{x} = 9$

31. $4\sqrt{x} + 5 = 41$

32. $\sqrt{x + 6} = 9$

33. $\sqrt{x - 7} = 12$

34. $8 + \sqrt{x} = 4$

35. $\sqrt{4x} = 12$

36. $\sqrt{5x + 15} = 0$

37. $\sqrt{5x} + 15 = 0$

38. $\sqrt{3x + 6} = \sqrt{2x - 4}$

39. $\sqrt{x + 8} = 2\sqrt{5}$

40. $5\left(\sqrt{3x} - 8\right) = 2\left(\sqrt{3x} - 2\right)$

41. Earlier you learned that the length of skid marks on the road can be used to estimate the speed of vehicles involved in crashes according to the formula $S = \sqrt{20D}$, where S is the speed in miles per hour and D is the length of the skid marks in feet. Solve that formula for D in order to predict how long the skid marks left behind would be if a car were travelling 60 miles per hour, 80 miles per hour, 100 miles per hour, or 120 miles per hour.

42. Your **Note to Self** about **solving radical equations** should include at least one example equation and a warning about the fact that some radical equations have no real solutions.

REVIEW

1. Graph the following pair of simultaneous inequalities:
   $y < 3x - 2$
   $y + x > 6$

2. A giant metal turnip is painted green, violet, and white in a ratio of 3 : 7 : 1. To the nearest whole percent, what percent of the turnip is green?

3. Solve the following equation:
   $$\frac{10}{x + 6} = \frac{2}{x - 10}$$

4. For all integers x, the made-up symbols V( ) and { } mean these things:

   $V(x) = \dfrac{x - 6}{2}$ and $\{x\} = x^2 + 1$

   Find $\{V(-10)\}$.

5. Graph the following pair of simultaneous inequalities:

$$y \leq -x^2 + 5$$
$$y > 2$$

6. Convert the equation y = -6x + 3 to two-intercept form and state what its x- and y-intercepts are.

Write equations to go with the following tables and tell what the basic shape of the graph would be (straight line, parabola, or hyperbola):

7.

x	-8	-6	-4	-2	2	4	6	8
y	-4	-5	-7	-13	11	5	3	2

8.

x	-13	-12	-11	-10	-9	-8	-7
y	-9	-4	-1	0	-1	-4	-9

9. Bronze-plated widgets sell for $11 apiece; hand-enameled widgets sell for $15 apiece. If you have four times as many bronze-plated widgets as hand-enameled widgets and the total value of your widget horde is $236, how many of each variety do you have?

10. A puzzle from BrainBashers.com:

Find a seven-digit number that describes itself in the following way: the first digit is the number of zeros in the number, the second digit is the number of ones, the third digit is the number of twos, and so on.

For example, a five-digit number that does this is 21200: it contains two zeros, one one, two twos, zero threes, and zero fours.

7 EXPONENTIAL FUNCTIONS

In this chapter you've dealt with a lot of expressions and a few single-variable equations, but so far almost no two-variable equations. Since it's been a little while, I want to remind you of the importance of two- (or more) variable equations. Single-variable equations are useful for problem-solving because they generally have a limited number of solutions (often only one). Two-variable equations aren't so much problem-solving tools as they are ways of describing and understanding the world. They have infinite solution pairs that can be represented graphically. They can be used to represent burning candles, the amount of energy in an object, the relationship between the speed of a car and the skid marks it leaves in a highway accident, or any number of other situations involving change.

At this point, you're quite familiar with three families of functions. (Remember that a function is a certain kind of two-variable equation that has only one value of the dependent variable for each value of the independent variable.) You know about linear functions such as $y = 3x - 6$ or $y = \frac{x}{10} + 16$, which have straight-line graphs.

You know about parabolic functions such as $y = x^2$ or $y = (x - 7)^2 + 6$, which have parabolic graphs. And you know about inverse functions, such as $y = \frac{6}{x}$ or $y = -\frac{10}{x + 2} - 5$, which have hyperbolic graphs. In this lesson, we'll look at an entirely new family of functions, which, if you pay attention to the lesson titles, you will not be surprised to learn are called *exponential functions*.

1. **Salvador Dalí, a famous artist with one of the most spectacular moustaches of all time, had a pet ocelot named Babou. They took a trip together aboard a luxury ocean liner. (This is all true. The next bit, not so much... at least I hope not.) Babou caught a rat and brought it to Salvador in his cabin, then returned to the hunt. Every hour after that, Babou managed to triple the number of rat carcasses in Salvador's cabin. Your job is to write an equation representing this increasingly unpleasant scenario, where y is the number of rat carcasses in the cabin and x is the number of hours that have passed since Babou caught that first rat. You might begin by making a table with a few pairs of values in it: when x is 0, y is 1; when x is 1, y is 3; when x is 2, y is 9; and so on.**

As I'm sure you will have anticipated, the equation you just wrote falls into the family of exponential functions: ones where the independent variable is part of the exponent in the equation.

2. **A flock of chimney swifts are holding a convention in the chimney of a local school. (Chimney swifts — *Chaetura pelagica* — are small, speedy, soot-colored birds, in case you didn't know.) At the start of the convention, there are 10 swifts in the chimney, and each minute the number of swifts doubles. Write an equation to represent this situation: make y the number of swifts in the chimney and x the number of minutes that has passed. (Hint: It will be similar to the equation from Problem 1.)**

All right, so those are two fictional examples of exponential functions. In a little while I'll give you some real-world examples. But first, what does the graph of an exponential function look like?

3. In the next problem, I'm going to ask you to graph $y = 2^x$. But to help you set up your graph in an efficient way, I'll ask you another question first: in the equation $y = 2^x$, can y be negative? Be sure to explain why or why not.

4. Now go ahead and graph $y = 2^x$ for the whole-number values of x from -4 to 4. Your graph will need to extend up to 16 on the y-axis. Use your answer from Problem 3 (or the table that you'll probably want to start by making) to figure out how far down the y-axis your graph should extend. Once you've got your nine points, you can go ahead and connect them with a smooth curve.

Notice that although the graph of **y = 2ˣ** bears some resemblance to the other graphs you're familiar with, it really belongs in its own family because it is different from any of those graphs. All exponential graphs have that same basic shape.

5. How is the graph of $y = 2^x$ similar to a parabola?

6. How is the graph of $y = 2^x$ similar to a hyperbola?

7. I'd like you to examine that resemblance to a parabola a little more closely. I think the best way to do so will be to graph $y = 2^x$ and $y = x^2$ on the same set of axes. You'll only need to graph the points where x is greater than or equal to zero, because those are the only similar parts of the graphs. Start by making tables of the values of $y = 2^x$ and $y = x^2$ for x-values from zero to 7. In order to compare them effectively on the graph, I'll ask you to use slightly weird scales: on the y-axis, one graph paper square should stand for two units, but on the x-axis, two graph paper squares should stand for one unit. Your y-axis should go up to 50. Go ahead and graph the equations, working carefully because of those weird scales and connecting your points with smooth curves. I recommend doing each curve in a different color because they cross each other. (Even with these scales, not all of the points you put in your table will fit on the graph, but that's okay.)

8. For what values of x are the y-values of $y = 2^x$ and $y = x^2$ equal to each other?

9. For what values of x on your graph are the y-values of $y = x^2$ greater than the y-values of $y = 2^x$?

10. For what x-values on your graph are the y-values of $y = 2^x$ greater than the y-values of $y = x^2$ — not just for the points on the chart, but for all the points represented by the curves?

11. For a very high value of x, would you expect x^2 or 2^x to be greater?

Because the **y-values** of exponential functions like **y = 2ˣ** get extremely large as **x** gets larger, people talk about things "growing exponentially" when they grow very rapidly. *Exponential growth* is a phrase that you'll hear on television or read in the newspaper fairly often. Indeed, it's often used to mean simply very, very fast growth — which I think is just fine, but you should be aware that the technical, mathematical definition of exponential growth is that the function describing the growth is an exponential one. There are numerous examples of situations that can be described using exponential functions.

One famous, fictional one has to do with a wise man, a king, and a chessboard. You may have heard it before. Some people claim that it is a Persian story, others that it is an Indian story, some that it involves grains of rice and others grains of wheat, some that the wise man is actually the god Krishna in disguise and some that the wise man is a wise woman or even a wise little girl. Regardless, the essence of the story is that the wise man earns a favor from a king. The wise man asks for some grains of rice every day: on the first day, he asks that one grain of rice be placed on the first square of a chessboard, on the second day that two grains be placed on the second square, four grains on the third square on the third day, then eight, sixteen, and so on, the number doubling every day. The king laughs at him and grants his request...

12. **The function that describes the number of grains the wise man receives is $y = 2^{x-1}$, where y is the number of grains and x is the number of days that have passed. (It's $y = 2^{x-1}$ because on the first day he gets 2^0, then 2^1 on the second day, and so on.) There are 64 squares on a chessboard. Calculate the number of grains that the wise man receives on the 64th day. (You'll need a scientific calculator for this. Depending on the calculator you use, the answer may be given in scientific notation, but you know how to interpret that! Calculators usually skip the "$\cdot 10$" part of scientific notation, so, for example, $4.44 \cdot 10^8$ would usually be shown as "4.44^8".)**

A grain of rice typically weighs about 20 mg (20 thousandths of a gram). What is the weight of the rice grains that the wise man receives on the 64th day? If he wants to rent some elephants to haul his rice home that day and each beast can pull 2,000 kg, how many elephants does he need? (Remember that 1 kg = 1,000,000 mg.)

And that's just the amount he would have received *on* the 64th day — it doesn't include the half of that amount he would have received on Day 63, or the half of that on Day 62, and so forth. This is a wonderful story for appreciating just how incredible exponential growth is. There are other good ones. For instance, I learned that if it were possible to fold a single piece of notebook paper in half **50** times (which it isn't, but feel free to test how many times you *can* do it), you'd end up with a piece of paper with a *very* tiny surface area that was thick enough to reach from here to the sun. (I won't ask you to do the calculations, but if you want to check this, it helps to know that a piece of paper is about $.1 \cdot 10^{-6}$ km thick.)

Those are some fantastical examples of exponential growth. What about real ones? Well, bacteria typically reproduce by splitting themselves in half, and hence doubling their population. This means that if you provide some bacteria with really pleasant living conditions — say, a petri dish full of nutrients — you can end up with a very large number of bacteria in a pretty short period of time. On a somewhat scarier note, viruses can spread exponentially through a population because each person (or animal) that is infected can infect multiple others. There are people who worry that the human population might be subject to exponential growth, though of course this would happen more slowly than it does in the case of bacteria because people take a lot longer to reproduce. Nuclear chain reactions grow exponentially — each atom that breaks apart releases energy that breaks other atoms apart, and so on. This is what enables nuclear power plants (and nuclear bombs) to produce so much energy from a relatively small amount of material.

As I've said before when talking about how functions can be used to describe the world, they often apply to a situation for only part of the time. Think back to the Crow and the Pitcher exercise from *Crocodiles & Coconuts*. This graph described what was going on:

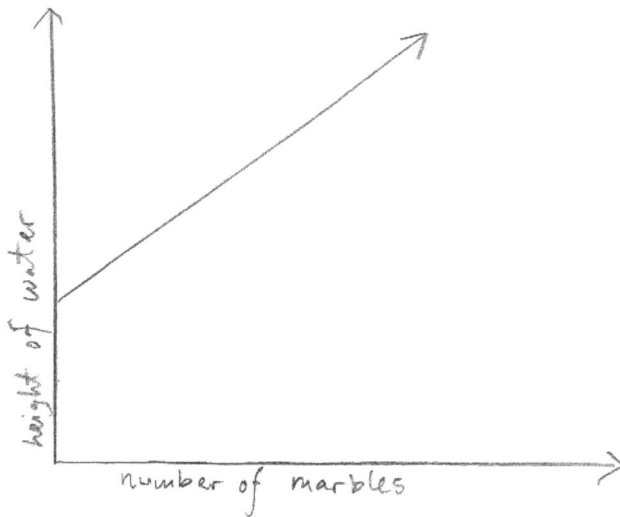

... but it was a little deceptive because it was really only true if you had an infinitely tall pitcher, so this graph was probably more accurate:

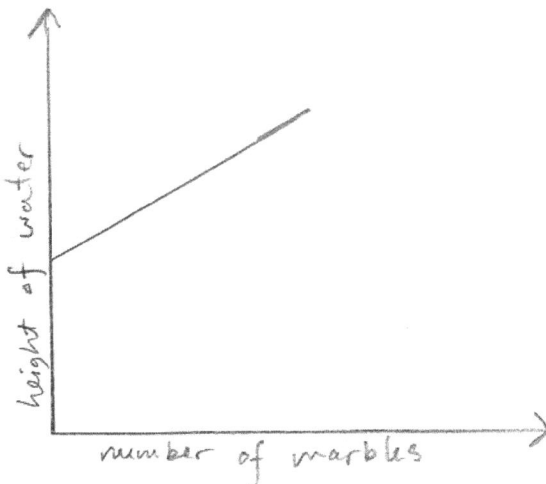

In the case of the Crow and the Pitcher, the height of the pitcher was a *limiting factor* — it limited how closely or for how long the linear function accurately described the situation.

13. **What are some limiting factors that might keep the growth of bacteria from being exponential in the long run?**

14. **What are some limiting factors that might keep the spread of a virus from being exponential in the long run?**

15. **What are some limiting factors that might keep the growth of the earth's human population from being exponential in the long run?**

Another example of exponential growth takes place in a bank account that earns compound interest, as long as you don't withdraw any money from it. "Compound interest" means that every time the bank pays you interest on the amount of money you have in the account, that payment gets added to the original amount — so that the next time you get paid interest, that interest is calculated based on the new amount in the account. For instance, if you open a savings account with **$100** and the account pays **5%** yearly interest, at the end of the first year you'd add **$5** to the account and have **$105**. At the end of the second year, you'd get **$5.25** (**5%** of **$105**) and have **$110.25**, and so on. Unfortunately, your bank account doesn't grow as fast as those bacteria.

16. The function that can be used to describe the bank account situation is $b = s(1 + r)^x$, where b is your current bank balance, s is the amount of money you started with, r is the interest rate, and x is the number of years that have passed. Use this formula to calculate how much money you'd have after 6 years in an account that you opened with $500 and that paid interest at a rate of 4%. (The only tricky part of this is to remember how to represent a percent as a decimal. You can go ahead and use a calculator for the actual calculations and round your answer to the nearest cent.)

17. How much money would you have if you didn't withdraw anything from the account for 64 years?

That's a decent amount of money, but...

18. Comparing the formula for your bank account in Problems 16 and 17 to the formula for the chessboard situation in Problem 12, why is it that the exponential growth of your bank account leads to so much less actual dollar growth than the exponential growth of rice grains on the chessboard?

In addition to exponential growth, you can have something called *exponential decay*, which is essentially the opposite of exponential growth. Here's an example of a graph of exponential decay:

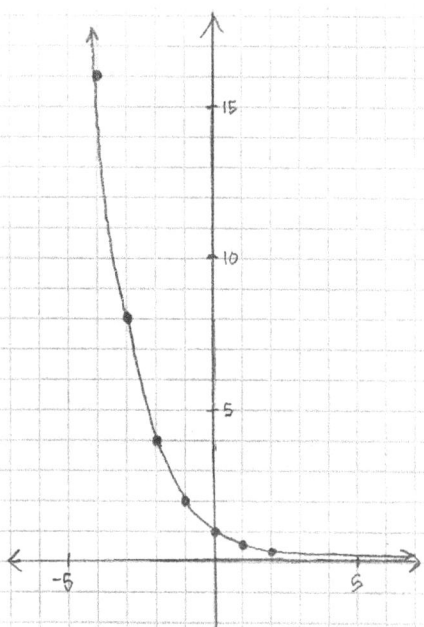

19. Determine the equation for the graph of exponential decay. Although it's a continuous curve, I've marked the points that I used to graph it and that I think will be most helpful in figuring out the equation. You might want to start by making a table of those points. There are two possible and equally correct versions of this equation, so when you've finished this problem, make sure you check in with your teacher and classmates so that you can see both versions.

Exponential decay occurs in any situation where something keeps decreasing by a fixed proportion. For example, if you have a sports tournament where half of the teams are eliminated in every round, you're seeing exponential decay. One of the most useful examples of exponential decay is called *radioactive half-life*. There are certain naturally occurring radioactive elements that break down predictably through exponential decay. The most famous example is carbon-14. Every living creature has carbon-14 in it, and once that creature dies, the carbon-14 starts to exponentially decay: half of it disappears about every **5,700** years. This means that if you have the remains of a living thing — say, a fossil — you can measure the amount of radioactive carbon-14 that remains in it and get an impressively accurate estimate of how old it is. (This method of estimating ages of fossils was developed by a man named Willard Libby and his colleagues in 1949. He won the Nobel Prize for his work.)

20. Write a **Note to Self** on the subject of **exponential equations**. Your Note should include examples of equations for exponential growth and exponential decay and sketches of graphs for both.

For the final problem set in this lesson, I'm going to ask you to categorize a set of equations just by looking at them. I think it will be extremely helpful to look back at the very last Note to Self that you wrote in *Crocodiles & Coconuts* (oddly enough, it may be called "Dr. Xavier Moosespackle's Fabulous Note-to-Self Extravaganza"), especially because at the end of the problem set I'll be asking you to add to that Note. As I've mentioned, the way to recognize an exponential function is that the independent variable appears in an exponent. Also, the graph of an exponential equation can be called an *exponential curve*.

For each of the following equations, state whether its graph would be a straight line, a parabola, a functional hyperbola, a non-functional curve, or an exponential curve. Assume that y is the dependent variable, x is the independent variable, and any other letters are constants.

21. $y = (3 - x)^2 + 16$

22. $y = 100 - \dfrac{22}{x + 2}$

23. $55 = y^2 - x^2$

24. $y = 23^x + .0006$

25. $y + x^2 - 13 = 0$

26. $c^{x - 2} + 3y = b$

27. $-18 = \dfrac{y}{5} + \dfrac{x}{4}$

28. $(3 - y)^2 = (3 - x)^2$

29. $y = mx + b$

30. $y = \dfrac{n}{m - x}$

31. $y = a^{-.007x}$

32. Add the information that you need to recognize that the graph of an equation is an exponential curve to that last Note to Self from *Crocodiles & Coconuts*.

REVIEW

Simplify the following expressions:

1. $(3.7 \cdot 10^{-6})(2.5 \cdot 10^{13})$

2. $\dfrac{7.1675 \cdot 10^{15}}{3.05 \cdot 10^{20}}$

Apply the appropriate Laws of Exponents to the following expressions:

3. $(-3x^2)^3$

4. $\dfrac{n^7}{n^{-6}}$

5. $m^7 n^3 m^{-2} n^5$

6. $(5^{10})(5^{-8})$

7. $\left(\dfrac{a^2}{b^3}\right)^5$

Estimate the value of the following:

8. $2\sqrt{10}$

9. $-3\sqrt{3}$

Express the following in simplest radical form:

10. $\sqrt{48}$

11. $\sqrt{162x^3}$

12. $\dfrac{\sqrt{45}}{6}$

13. $\dfrac{\sqrt{5}}{\sqrt{3x}}$

14. Simplify the expression $3\sqrt{7} - 10\sqrt{7}$.

If possible, simplify the following expressions:

15. $3\sqrt{7} - 10\sqrt{7}$

16. $\sqrt{28} + \sqrt{63}$

17. Apply the Distributive Rule to the expression $3x(\sqrt{5} + \sqrt{3})$.

18. Solve the equation $\sqrt{x + 5} = 6$.

19. Solve the equation $3\sqrt{x} - 10 = \dfrac{\sqrt{x}}{2}$.

20. Solve the equation $3\sqrt{x} + 10 = 4$.

21. For the equation $y = 3^{-x}$, find the values of y when x is 3, 1, 0, -1, and -3.

2

INTRODUCTION TO POLYNOMIALS

1 WHAT'S A POLYNOMIAL & WHAT'S IT FOR?

For much of the rest of this textbook, you'll spend a lot of time working with a new mathematical subject: *polynomials*. So what are polynomials? Actually, they're just mathematical expressions like ones you've seen before. Really, *polynomial* is a name that represents a new way of thinking about these sorts of expressions. Polynomials come in many varieties; *monomials*, *binomials*, and *trinomials*, for example, are all types of polynomial. You can also use words like *quatrinomial* or *quintnomial* if you like, but most people don't. In general, anything beyond the level of a trinomial is just called a polynomial.

I think that the best way to start understanding what they are is to look at some examples. Here is a list of things that can be called monomials:

$7x$
$3x^4$
x^3
$16x^2$
$8x^2$

(I've only used the variable x, which is a pretty common choice, but actually any variable is fine in a polynomial.)

Here are some binomials:

$4x^2 + 2x$
$x^3 - x$
$3x^5 + 7$
$14x^4 - 20x$
$x^{17} + 2$

... and here are some trinomials:

$4x^3 + 17^x - 9$
$-5x^5 + 2x^3 + 4$
$x^2 + 3x + 2$

... and here are some other polynomials:

$85x^8 - 13x^4 + 12x$
$2x^8 + 2x^7 + 2x^6 + 2x^5 + 2x^4 + 2x^3 + 2x^2 + 2x + 2$
$5x^6 + x^4 - 13x^2 + 1$
$-3x^6 + 8x^5 - 9x^3 + x^2 - 4$

1. **Based on those seventeen examples, what do you think polynomials are — that is, are there any characteristics that all seventeen examples seem to have in common?**

2. Based on those seventeen examples, what is it that makes a monomial a monomial, a binomial a binomial, and a trinomial a trinomial, as opposed to the other varieties of polynomial? (It may help you to think of other things you know that have the prefixes *mono*, *bi*, *tri*, or *poly* — or to look those prefixes up if you don't know them. Three of them are Greek and one is Latin.)

Let's look at monomials first, since they are the simplest polynomials. You've seen expressions that are monomials over and over again as you've studied algebra — it's just that up to this point you haven't called them that. As you doubtless figured out, monomials are polynomials that have just one part — it's usually called a *term*. Here's one:

$2x^5$

There are two pieces of vocabulary that go with monomials: in the monomial $2x^5$, the **2** is called the *coefficient* and the **5** is called the *degree*. (You can say that it's a *fifth-degree* monomial if you like.)

Name the coefficient and degree of each of the following monomials:

3. $7x^4$

4. $19x^2$

5. $2x^6$

6. $5x^5$

7. $\frac{1}{2}x^3$ (Actually, you'll rarely see monomials with fractional coefficients in this book or elsewhere, but I want you to realize that there's nothing wrong with them.)

8. $\frac{x^5}{3}$ (Remember that division is the same as multiplying by a fraction.)

9. x^3 (Yes, there is a coefficient for this one, and it's not zero.)

10. $5x$ (Yes, this one has a degree. Look back at your Notes to Self having to do with exponents if you have trouble figuring it out.)

One more thing to know about monomials before we move on to bi-, tri-, and polynomials. I imagine that in Problem 1 above, you might have said that all polynomials include an **x**, and that would have been a perfectly valid conclusion based on the examples I gave you. But here's another example of a monomial:

12

I'll bet that through all those years of working on math in elementary school you never realized that the numbers you were working with were actually monomials.

11. What justification do mathematicians have for thinking of the number 12 (or any other number) as a monomial? To answer this question, you need to realize that mathematicians (when they're working with polynomials, anyhow) think of 12 as having a *hidden* x-component — that is, they think of 12 as a coefficient being multiplied by x raised to a certain degree. So the question really becomes, what degree could x be raised to that would cause it effectively to disappear? Here are a couple of hints: 1) Definitely look back over your work with exponents, or at least think carefully about that work. 2) In a way, the answer to this question has to do with why x^3 really does have a coefficient, just a hidden one.

Now let's go on to look at bi- and polynomials. (Even though binomials are given a special name, we're going to consider them along with all other polynomials with more than one term.) So, here's a polynomial with two terms (or a binomial):

$7x^4 - 2x$

Here's a three-term polynomial (a trinomial):

$3x^6 - x^5 + 2x^2$

Here's a polynomial with four terms:

$-7x^4 + 8x^3 - 6x + 4$

And here's one with ten terms:

$2x^{13} + x^{11} - x^{10} + 3x^9 - 4x^7 + 2x^5 + 13x^4 - 7x^3 + 9x - 137$

I hope it's now crystal clear what I mean by the "terms" of a polynomial. I hope it's also clear what I mean when I say, for example, that the second term of that ten-term polynomial is x^{11}.

Here's another important piece of vocabulary: when you have a polynomial like, for example, $3x^2 + 6x + 5$, the **5** is called the *constant* of that polynomial. It's called a constant because it stays the same even when **x** changes.

There are a few important things to notice about polynomials. The first has to do with the standard order in which the terms of a polynomial are written.

12. Look back at all of the polynomials I've written so far in this lesson. They all follow a rule for the order of the terms that has to do with the degrees of the terms. What is the rule?

Actually, it would probably be better to call that a "convention" rather than a "rule." As you know very well, $5x^4 + 3x^3$ is equal to $3x^3 + 5x^4$, and there's no law against writing it the second way, but in this textbook (and most others) you'll generally see it written the first way for reasons that you'll explore in just a moment.

(By the way, while we're on the subject of this convention for writing polynomials from highest-degree term to lowest-degree term, I should say that I don't know of any reason why the terms of a polynomial couldn't have negative degrees, but I've never seen one with negative degrees and I won't ask you to work with any in this book. If you were to work with one, I suppose the convention would still hold, so you'd write it like this: $2x^2 + 3x + 7 + 2x^{-1} + x^{-2}$.)

Even though each term of a polynomial has its own degree, the polynomial as a whole is also said to have a degree, which is just the degree of its highest-degree term. So, this is a third-degree polynomial:

$5x^3 + 2x^2 - x + 7$

This is a fifth-degree polynomial:

$17x^5 - 4x + 17$

This is an eighth-degree polynomial:

$x^8 - 7x^5 + 13x^3 - 2x$

... and so on.

So why should a polynomial be written in order of degrees and named for its highest-degree term?

13. In order to answer that question, consider the polynomial $x^3 + x^2 + x + 1$. As you know, x can have any value, and I'm going to ask you to consider the values of the terms of that polynomial for several possible values of x. I think the best thing to do is to make a chart like the one below in your notebook and finish filling it out. I have two pieces of advice. First, make your boxes pretty big, as I've done — when x is 1,000,000, x^3 will be quite large. Second, don't get confused when you ask yourself, "What is the value of 1 when x is 10?" and so on. The answer to that question really is as obvious as you think.

x^3	x^2	x	1
1	1	1	1
	100	10	
		1,000	
		10,000	
		1,000,000	

14. Examining your chart, you'll see first off that when x is equal to 1, each of the terms of $x^3 + x^2 + x + 1$ is equal to 1. However, that changes as x gets bigger. Based on what you see in your chart, why do you think the x^3 gets written first and why do you think $x^3 + x^2 + x + 1$ is referred to as a third-degree polynomial?

I hope you were just reminded of how powerful exponents are. In the polynomial ($x^3 + x^2 + x + 1$), if x were **1,000,000**, the third term (**x**) would be equal to **1,000,000** — a pretty big number. But the first term would be equal to **1,000,000,000,000,000,000** — a much, much larger number. In other words, the term with the highest degree tends to dominate the value of a polynomial, especially as the value of the variable gets very large, which is precisely why a polynomial is named for its highest degree and also why the terms are written in descending order of degree.

(Notice that, in the last paragraph, I put the polynomial ($x^3 + x^2 + x + 1$) in parentheses. I will sometimes do that if I think it helps make it clear that I'm referring to the whole polynomial and not just its terms. I'll also sometimes put equations in parentheses. I won't use parentheses if I don't think there's any risk of confusion.)

The next thing that I'd like you to notice is that none of the polynomials I've written in this lesson can be simplified any further than they already are. This is review, but it's very important. You need to be absolutely sure of the fact that this polynomial cannot be simplified:

$$3x^3 + 2x^2$$

... but this one can:

$$3x^3 + 2x^3$$

The reason for this is that x^3 and x^2 are different from each other. As you just saw in Problem 13, if **x** is a large number, they can be radically different — say **1,000,000,000,000** versus **1,000,000,000,000,000,000**. On the other hand, x^3 and x^3 are the same thing, no matter what the value of **x** is.

Trying to add $3x^3$ and $2x^2$ is an apples and wingnuts situation. Adding $3x^3$ and $2x^3$, on the other hand, is perfectly simple. Three of anything plus two of anything is five of those things: $3x^3 + 2x^3 = 5x^3$.

Whenever you see a polynomial, you should get used to making sure that its terms are written in descending order of degree and that it's as simplified as it can be — that all terms of the same degree have been added to or subtracted from one another.

Simplify the following polynomials and write them properly.

15. $2x^4 + x^7 - 3x + x^4 + 5x^8$

16. $10x + 15 - 2x^2 - 3x + 22$

17. $14x^3 + 14x^2 + 14x - 6x + 3x^2 - 3x^3$

These next problems are a review of your exponent work that will come in handy. Rewrite each of the following as a monomial:

18. $2x \cdot 8x$

19. $3x^2 \cdot 3x^3$

20. $3x^4 \cdot 5x^3 \cdot x$

21. $6x^2 \cdot 2x^2 \cdot 7$

22. $12x(-x)$

23. $12x(-12)$

24. $x^5 \cdot x^5 \cdot x^5$

25. $(3x^5)^2$

26. $(-2x^4)^3$

For these next ones, you'll need to use the Distributive Rule. You'll end up with binomials. Simplify the following expressions:

27. $6(3x - 4)$

28. $x(3x - 4)$

29. $(-2x)(-2x + 5)$

30. $3x^2(x^5 + x^3)$

There's one more thing that I'd like you to realize about polynomials for right now — or perhaps it's a way of thinking about polynomials that I'd like you to try to adopt. Here's a polynomial:

$-7x^4 + 3x^3 - x^2 + x - 27$

Notice that the first term of this polynomial is negative. That's the way of thinking that I'd like to encourage you to make a habit of: it's not that the first term is being subtracted from anything — it's that it's negative. And that goes for the third and fifth terms as well. As you know, subtraction is the same as adding a negative, so this isn't a new idea for you, but believe me that the more you get into the habit of thinking of positive and negative terms of polynomials rather than about the terms being added or subtracted, the better off you'll be when you work with polynomials. In fact, to remind you to practice this habit, I'm going to start leaving a space between the negative sign and the first term of the polynomial, even though that's not the way you're used to seeing it. So, when you think about that polynomial:

$- 7x^4 + 3x^3 - x^2 + x - 27$

... think of the second and fourth terms as positive and the first, third, and fifth terms as negative.

So, now that you know the essentials of what polynomials are, I hope that you may be asking yourself, "So what? Why should I suddenly start thinking of these expressions as 'polynomials'? What does this have to do with any of the algebra I've been learning up until now?" You'll be working with polynomials throughout the rest of this book and at first you'll just be sort of messing around with them — learning to manipulate them. So I'll try to give you a sense now of why you should bother doing that.

In my view, all of algebra — or at least all of the algebra that you'll cover in this series of three books — has three main purposes, or main ideas, or main threads... whatever you want to call them.

The first purpose of algebra is that it can be used to solve problems. This is essentially what you practiced with all the single-variable equations in *Jousting Armadillos*. When you encounter an equation such as, say, **3x + 17 = 50**, you can figure out what value of **x** makes the equation true. Even more importantly, if you encounter certain kinds of problem-solving situations and you can figure out what you want the variable to stand for, you can write your own equation and use it to find a solution.

The second purpose of algebra (and my own personal nomination for the most important) is to describe and understand the world. Here I'm talking about two- (or more) variable equations — the work you did in *Crocodiles & Coconuts*. When you encounter an equation such as **y = 3x + 17**, you understand that there are infinite possible solutions, but that the pairs of numbers that make up those solutions have a very specific relationship to each other. You know how to recognize and sketch the graphs of equations belonging to a number of families, and you know that some of those equations represent real relationships in the world.

The third purpose of algebra, the one with which you've had the least experience, is that it can be used to prove certain mathematical facts.

The use of polynomials will extend your abilities in all three of these areas.

Right now, you could find the solution to **3x + 17 = 50** very swiftly. You could probably do it in your head while you're reading this.

31. **Go ahead and solve it, just to show that you can do it. What the heck.**

But if I asked you to find the solution(s) to this equation:

$$x^3 - 3x^2 - 10x = 0$$

… you would have a much harder time. In fact, the best you could probably do is to guess and check. You could probably get one of the solutions (there are three) pretty quickly: the equation is true when **x** is equal to zero.

32. **Just to underline my point, go ahead and try to find the other two solutions to ($x^3 - 3x^2 - 10x = 0$). Spend at least two minutes (unless you find them in less than two) but not more than five minutes trying to do this.**

Maybe you found them and maybe you didn't, but I hope you appreciate that guessing and checking is probably not a very efficient method for this kind of situation. The other two solutions are **5** and **-2**.

33. **Check to see that all three solutions — x = 0, x = 5 and x = -2 — really do make the equation (x^3 - $3x^2$ - 10x = 0) true.**

The work that you're going to do with polynomials will enable you to find those three solutions without guessing and checking.

In a similar fashion, if I asked you to graph the equation **y = 3x + 17**, you could practically do it in your sleep. In fact, I'll bet you're already picturing it in your head: a straight line with a **y-intercept** of **17** and a slope of **3**. If, on the other hand, I asked you to graph the equation (**y = x^3 - $3x^2$ - 10x**), you would, I expect, be pretty much stumped.

As it happens, a sketch of that graph would look like this:

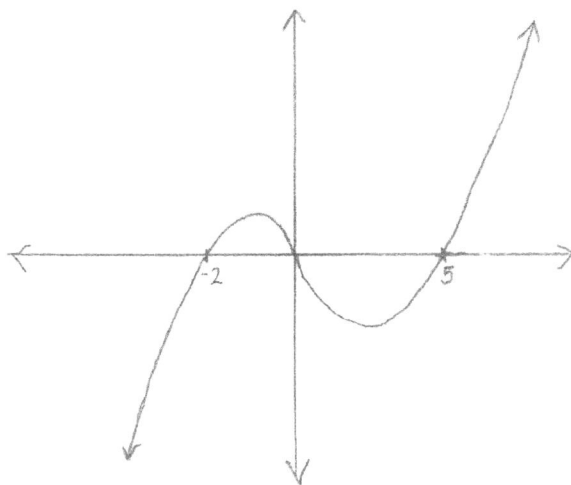

Your work with polynomials will enable you to draw the graphs of equations like (**y = x^3 - $3x^2$ - 10x**).

34. **What's the relationship between the points where the graph of (y = x^3 - $3x^2$ - 10x) crosses the x-axis and the solutions of the equation (x^3 - $3x^2$ - 10x = 0)?**

Interesting coincidence, no? We'll explore that coincidence in the next chapter.

Your work with polynomials will also allow you to begin exploring algebra's third purpose of proving certain mathematical facts.

35. **It's time for a *Note to Self*. As you can well imagine, many of your Notes for this book will contain the word *polynomials* in the title. A good name for this one might be *How to Write Polynomials,* or maybe *Writing Polynomials*. It should contain several examples of polynomials and show how to write them properly. It should explain what coefficients and degrees of terms are and what the degree of a whole polynomial is. It should also show examples of simplifying polynomials by combining terms of the same degree.**

REVIEW

1. Make approximate copies of the two graphs below in your notebook and explain why one could be the graph of a function and the other could not.

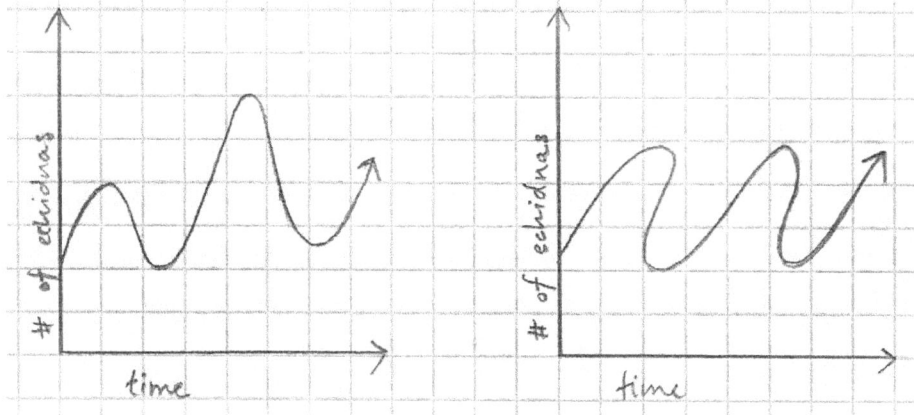

2. If Gwendolen can weed 2 standard garden plots in 3 hours and Cecily can weed 1 standard garden plot in 4 hours, how long will it take them to weed 2 standard garden plots working together?

3. Graph the equation $x^2 + y^2 = 25$.

4. Find the slope of a straight line that passes through the points (-9, -11) and (-6, 10).

5. Solve the equation $\frac{-5x - 1}{2} - 5 = \frac{3x - 1}{5} + 4$.

6. Convert the expression $\sqrt{150x^7}$ to simplest radical form.

7. Represent the inequality $-5 \leq x < 3$ on a number line.

8. If there are $4.2 \cdot 10^{15}$ bacteria in a 2-liter container, how many bacteria are there per milliliter?

9. Linus and an armadillo each went boating. The armadillo traveled 10 nautical miles an hour faster than Linus did. The armadillo sailed for five leisurely, fun-filled hours and Linus rowed for an exhausting 11 hours, but at the end of the day the armadillo had still traveled 8 nautical miles farther than Linus had. How fast were they each going?

10. Here's another puzzle adapted from one of Norman Willis's.

You are faced with three boxes and you know that two of them contain poisonous snakes and one contains doughnuts. They are labeled as follows:

| Exactly two of these labels are false. | Open this box for a yummy treat! | This box contains doughnuts. |

If you have to open one box, which should it be and why?

2 ADDING & SUBTRACTING POLYNOMIALS

Adding and subtracting polynomials is so much like the simplifying work that you did in the last lesson that it's almost not worth having as a lesson on its own, but I didn't want to overload Lesson 1.

Of course, at the end of Lesson 1, didn't I say something about how it was much better to think of polynomials as having positive and negative terms rather than adding or subtracting? That's absolutely right when we're talking about the terms of a single polynomial. When you work with the polynomial ($3x^5 - 6x^3 + 2x - 7$), you should think of the first and third terms as positive and the second and fourth terms as negative. However, it is possible to add one entire polynomial to another entire polynomial or to subtract one entire polynomial from another entire polynomial.

1. **We'll start with a simple case: adding ($3x^3 + x^2 + 7x + 15$) to ($x^3 + 5x^2 + 3x + 6$). Honestly, I think you can figure out how to add those two polynomials without any help from me. Here's your one hint: The answer will be another polynomial, also with four terms.**

Basically, there are two ways to think of it.

If you like, you can string all of the terms of the two polynomials together as a single polynomial, grouping terms that have the same degree, like this:

$3x^3 + x^3 + x^2 + 5x^2 + 7x + 3x + 15 + 6$

Then you simplify that polynomial according to the method from the last lesson and get:

$4x^3 + 6x^2 + 10x + 21$

If you prefer, you can write the two polynomials one on top of the other as if they were a traditional addition problem — remembering to put terms that have the same degree in the same column — like this:

$3x^3 + x^2 + 7x + 15$
$\underline{x^3 + 5x^2 + 3x + 6}$

And then add each column, once again getting:

$4x^3 + 6x^2 + 10x + 21$

Either method is just fine. Generally speaking, I do the actual steps of addition in my head, but you should always write the original problem in your notebook anyhow — this makes it much easier for you, your classmates, and your teachers to check your work. If you prefer to use the column form, you should just go ahead and write the problem down in that form.

2. Add $(3x^6 + 2x^4 + x^3 + 5x^2 + 7)$ to $(6x^6 + 2x^4 + 13x^3 + 7x^2 + 7)$.

3. Add $(12x^3 + 5x^2 + 19x + 4)$ to $(5x^3 + 16x^2 + x + 14)$.

4. Add $(6x^{35} + 4x^{20} + 9x^{15} + x^{10})$ to $(7x^{35} + 9x^{20} + 5x^{15} + 100x^{10})$.

There are just a couple of things to watch out for when adding polynomials. The first is that sometimes the two polynomials don't have terms whose degrees match up so nicely. Again, just keep in mind what you know about simplifying polynomials.

5. Add $(5x^4 + 3x^2)$ to $(2x^3 + 4x + 12)$.

As you just saw, it's possible for two polynomials to have no terms of the same degree. In that case, when you add them, you just basically list the terms of the two polynomials in order of descending degree.

Often, two polynomials will share some terms of the same degree but not others.

If your method of adding is to turn the two polynomials into a single one and then add, be careful not to simplify anything that can't be simplified. If you prefer to think of adding polynomials in terms of traditional column-style addition, you should leave blank columns or put in terms with zero coefficients (which amounts to the same thing). So, if you were adding $(5x^6 + 7x^3 + 3x^2 + 4x + 9)$ and $(4x^4 + 3x^3 + 8x^2 + 7)$ you could arrange it like this:

$$5x^6 \quad\;\; + 7x^3 + 3x^2 + 4x + 9$$
$$\underline{\quad\;\; 4x^4 + 3x^3 + 8x^2 \quad\quad\; + 7}$$

… or like this:

$$5x^6 + 0x^4 + 7x^3 + 3x^2 + 4x + 9$$
$$\underline{0x^6 + 4x^4 + 3x^3 + 8x^2 + 0x + 7}$$

6. Add $(6x^5 + 2x^3 + 8x^2 + 3)$ to $(3x^4 + 9x^3 + 5x + 10)$.

The other thing to be aware of is that the terms of the polynomials won't always be positive, as they have been in all of the previous problems. Just remember that when you add a negative it's like subtracting.

7. Add $(6x^3 - 5x^2 + 4x + 3)$ to $(7x^3 + 2x^2 - 3x - 5)$.

8. Add $(12x^5 + 18x^3 - 4x^2 - 14x + 3)$ to $(- 4x^4 - 4x^3 + 10x + 16)$.

9. Add $(4x^9 - x^7 + 2x^5 - 3x^4 + 10x^2 - x + 10)$ to $(- 4x^9 + 3x^8 + 2x^7 - x^6 - 3x^5 + 10x^4 + 5x - 12)$.

10. Add $(7x^5 - 6x^4 + 3x^3 - 2x^2 + 15x + 100)$ to $(- 7x^5 + 6x^4 - 3x^3 + 2x^2 - 15x - 99)$.

11. Add $(14x^{28} - 10x^{22} + 10x^{18} - 3x^{12} + 9x^6 - 18)$ to $(- x^{28} + 5x^{25} - 10x^{18} + x^{13} - 6x^{12} + 7x^8 - 22)$.

Of course, there's no reason that you can't add more than two polynomials…

12. Add $(7x^5 - 4x^4 + x^3 + 10x^2 - 6x + 5)$ to $(-3x^5 + 13x^4 + 4x^3 - 3x^2 + 9x - 22)$ to $(8x^5 + x^4 - 3x^3 - 7x^2 + x + 16)$.

13. Add $(3x^3 + 4x^2 - 6x + 8)$ to $(4x^3 + 5x^2 - 2x - 4)$ to $(-8x^3 - 2x + 6)$ to $(x^3 - 10x^2 - 10)$.

Subtracting polynomials is slightly trickier than adding them, but only slightly. The one thing that you have to keep in mind is that you are always subtracting an entire polynomial from another. For instance, suppose you were to subtract $(4x^3 + 6x^2 + 5x + 9)$ from $(7x^3 + 8x^2 + 3x + 9)$. I'm careful to think of it like this:

$(7x^3 + 8x^2 + 3x + 9) - (4x^3 + 6x^2 + 5x + 9)$

... or like this:

$$\begin{array}{l} (7x^3 + 8x^2 + 3x + 9) \\ -(4x^3 + 6x^2 + 5x + 9) \\ \hline \end{array}$$

However you picture it, it works like this: subtract $4x^3$ from $7x^3$ and you get $3x^3$ (no problem); $8x^2$ minus $6x^2$ (because you are subtracting each piece of the second polynomial) is $2x^2$; subtract $5x$ from $3x$ and you get $-2x$; 9 minus 9 is zero. So the polynomial that you get as an answer is $(3x^3 + 2x^2 - 2x)$.

The one common mistake I've seen students make with these next problems is to forget what it means to subtract something from something else. Subtracting 6 from 10 means $10 - 6$.

14. Subtract $(7x^6 + x^5 + 3x^3 + 2x + 13)$ from $(6x^6 + 3x^5 + x^3 + 9x + 20)$.

15. Subtract $(4x^4 + x^3 + 10x^2 + 3x + 3)$ from $(6x^4 + x^2 + 7x + 15)$.
(Be conscious about what you get when you subtract x^3 from zero.)

16. Subtract $(6x^{12} + 8x^9 + 11x^5 + 3x^3 + 7)$ from $(4x^{12} + x^{10} + 8x^9 + 2x^6 + 3x^5 + x^3 + 15)$.

The situation gets more complicated, as you might expect, when the signs of the terms of the polynomials that you're working with are not all positive. For example, suppose you want to subtract $(7x^3 - 3x^2 + 4x - 6)$ from $(x^3 + 3x^2 - 6x - 5)$. You'd set it up something like this:

$$\begin{array}{l} (x^3 + 3x^2 - 6x - 5) \\ -(7x^3 - 3x^2 + 4x - 6) \\ \hline \end{array}$$

... and then you'd do it just the way you did the other ones, being very careful about subtracting negatives: x^3 minus $7x^3$ is $-6x^3$; subtracting a negative is the same as adding, so $3x^2$ minus $-3x^2$ is $6x^2$; subtract $4x$ from $-6x$ and you get $-10x$; -5 minus -6 is 1. So the answer is $(-6x^3 + 6x^2 - 10x + 1)$.

17. Subtract $(5x^4 - 6x^3 + x^2 + 4x - 3)$ from $(x^4 + 4x^3 - x^2 + 7x + 6)$.

18. Subtract $(- 4x^6 + 4x^4 - 3x^3 + x^2 - 8x + 16)$ from $(- x^6 + 5x^5 - 4x^4 - x^3 + 14x + 2)$.

19. Subtract $(15x^5 + 7x^4 - 3x^3 + 50x - 1)$ from $(x^6 + 17x^5 - 30x^4 + x^3 + 88)$.

20. Subtract $(- 4x^3 + 6x^2 + 18x - 14)$ from $(6x^3 + 6x^2 - 20x - 13)$.

21. Subtract $(7x^8 - 6x^5 + 4x^3 - 9x^2 + 3x + 5)$ from $(- 2x^8 + 4x^7 + 6x^5 - 3x^3 + x^2 + 8)$.

22. Add $(6x^3 + 7x^2 - 4x + 19)$ to $(7x^3 - 5x^2 + 10x + 13)$ and then subtract $(x^3 - 5x^2 + 11x + 7)$. (I wrote, "and then subtract," but actually, if you think in terms of adding negatives, it doesn't matter what order you go in.)

I'm going to throw in a few polynomials that have more than one variable in them. Although the majority of the polynomials that you'll work with in this book will have only one variable, there's no law saying that you can't have a multi-variable polynomial. There's also no really strict way of writing them. For example, $(x^2 + 3x + 2y^3 + y^2 + 6)$ is just as good as $(2y^3 + y^2 + x^2 + 3x + 6)$ or any other combination. (Usually the constant — **6**, in this case — goes last.) Just don't fool yourself into thinking that terms like x^2 and y^2 can be simplified further because they're both raised to the same degree. $(x^2 + 3x + 2y^3 + y^2 + 6)$ is as simple as that polynomial gets.

23. Add $(4x^3 + 5y^3 + 2x^2 - 7y^2 + 8)$ to $(x^3 - 3y^3 + 6x^2 + 2y^2 + y - 5)$.

24. Add $(4x^4 - x^2 + 3y^3 - 5y + 9)$ to $(6x^3 + x^2 - 5y^3 + 4y + 2x + 10)$.

25. Subtract $(7x^3 + 7y^3 + 4x^2 - 3y^2 + 6x - y + 7)$ from $(6x^3 - y^3 + 10x^2 + 3y^2 - 9x - y + 18)$.

26. Subtract $(4y^2 + 9y + 6x^2 - x - 100)$ from $(3y^3 + y^2 - 7y + 10x^2 + 40x + 2)$.

27. Add $(x^2 + 3xy - y^2)$ to $(4x^2 + 2xy + 3y^2)$. (The question you should ask in simplifying this one is, "Can 3xy be added to 2xy?" And I'll ask you, "Is this a case of adding apples and wingnuts or adding apples and apples?")

28. Write a **Note to Self** on the subject of **adding and subtracting polynomials**. Explain the techniques you learned in this lesson and give a couple of examples.

In my opinion, this was one of the easier, shorter lessons you've had in a while. Adding and subtracting polynomials just isn't that tricky. There's one more subject I'd like to touch on before we move forward, and that's this idea of using algebra to prove things.

I haven't really put any emphasis on this so far in this series of textbooks, but from a certain perspective you've been proving things all along. Take, for example, this little bit of simplifying that you might have done back in *Jousting Armadillos*:

$3x + 2x = 5x$

Translated into English, that statement might read, "Three of an unknown number plus two of that same number is the same thing as five of that number." Or another way of translating it might be, "Three of any number plus two of that same number is the same thing as five of that number." If you translate it that second way, it kind of puts a different spin on it. Because if you really believe the algebraic statement "**3x + 2x = 5x**", then you believe that it's true for any possible number in the universe. I don't care if we're talking about whole numbers or integers or rational real numbers or imaginary numbers. Three of any number plus two of that same number is the same thing as five of that number. So, in a sense, when you write "**3x + 2x = 5x**", you could say that you're proving something.

29. **Look back at Problem 27. If your work on that problem was correct, then it has to be true for any two numbers, x and y. Choose three pairs of x and y — any pairs that you want. Plug those two numbers into the expressions ($x^2 + 3xy - y^2$) and ($4x^2 + 2xy + 3y^2$) and add the results. Then plug those same three pairs into the expression ($5x^2 + 5xy + 2y^2$) and see what you get. It might make sense to keep track of your results in a table like this one:**

x	y	$x^2 + 3xy - y^2$	$4x^2 + 2xy + 3y^2$	$(x^2 + 3xy - y^2)$ $+(4x^2 + 2xy + 3y^2)$	$5x^2 + 5xy + 2y^2$

Okay, you haven't proved anything terribly profound or interesting yet. But here's one last problem for the lesson:

30. **This problem is from Harold Jacobs's *Elementary Algebra* and it has to do with adding polynomials. Think of a two-digit number where the ones' digit is greater than the tens' digit. Multiply the difference between the two digits by 9 and add the result to your original number. Try this for two or three different numbers. What always seems to happen? Now, use algebra to *prove* that this will always happen. Here's your hint: If you let x and y represent the tens' and ones' digits of a two-digit number, then the number itself can be expressed as 10x + y. I'd like you to notice that you're using algebra to prove *deductively* what you just saw *inductively* with your three examples. (This is the first time you've really done something quite like this, so it wouldn't surprise me if you needed support from your math partners and/or teachers. Regardless, I'd like you to make sure you discuss it with a teacher when you're done.)**

REVIEW

1. **Graph the inequality** $y \geq \dfrac{6}{x}$.

   (You've never graphed an inverse inequality before, but the skills involved are all ones that you've learned.)

2. Solve the equation $3\sqrt{x+6} + 10 = 16$.

3. Express $\dfrac{1}{64}$ as a power of 4.

4. Simplify the following expression:

$$\frac{90}{x^3 y^6} \cdot \frac{xy^3}{105}$$

5. Simplify the following expression and state which values of x and y are prohibited:

$$\frac{0x+y}{y} + \frac{0}{x-y}$$

6. Write an equation to go with the following table and state what the basic shape of its graph would be:

x	0	1	2	3	4
y	24	35	48	63	80

7. Express the following in simplest radical form:

$$\frac{\sqrt{50x^3}}{\sqrt{5}}$$

8. Solve the following equation:

$$\frac{y}{3} + \frac{17}{9} = 6y$$

9. Four people are gathered together. The second is three years older than the first, the third is four years older than the second, the fourth is five years older than the third, and the oldest is five times as old as the youngest. How old are they?

10. A puzzle from BrainBashers.com:

I'm sick and my doctor has prescribed two medications for me: Boxiloxiloxican and Ambroxilite. It's vitally important that I take one tablet of each every night. Tonight, I'd taken out a tablet of Boxiloxiloxican and had it in the palm of my hand, but when I tipped the jar of Ambroxilite into that same hand, two tablets fell out. The problem is that the tablets all look exactly alike and now I can't tell which of the three pills in my hand is which. They are quite expensive and so I'm loath to just throw them away and start over. How can I resolve this dilemma?

3 MULTIPLYING POLYNOMIALS

You've already dealt with the two simplest cases of multiplying polynomials. The very simplest case is multiplying a monomial by a monomial, like this:

4x · 6x

... or this:

5(3x²)

... or even (if you want to look at these as monomials) this:

7 · 5

Those should be pretty easy for you. The second-simplest case of multiplying polynomials is multiplying a monomial by a binomial. Here's an example of that:

3(x + 7)

You started working with expressions like that one way back in Chapter 2 of *Jousting Armadillos*. At that point I called it "applying the Distributive Property." And that's exactly what it is. In fact, as you learned in *Jousting Armadillos*, almost every time you do a multiplication problem you are applying the Distributive Property. For example, when you multiply **6 · 28**, you use the Distributive Property: you multiply **6** by **8** to get **48**, you multiply **6** by **20** to get **120**, and you add **48** to **120** to get **168**. That's almost exactly the same process as doing this:

3(x + 7) = 3x + 21

Actually, multiplying a monomial by any size polynomial works pretty much the same as multiplying a monomial by a binomial: just be sure that you multiply the monomial by *each term* of the polynomial. In every case, you're applying the Distributive Property. Here are a few to try, including a couple of classic monomial-by-binomials to warm up with:

1. 7(x - 6)

2. 2x(x + 8)

3. x(y - 4x)

4. 6(x² + 3x + 4)

5. 5(3x³ + 4x - 7)

6. - 7(- 4x² + 7x - 10)

7. - 3x(6x² + x - 2)

8. 4x²(- 2x² + 8x - 8)

9. y(3x² + 2x + 1)

10. 6(4x⁵ + 3x⁴ - x³ + 4x² - 12)

11. **Based on the problems you just did, when you multiply a monomial by a polynomial, how many terms does your polynomial answer seem to have? In other words, when you multiply a monomial by a binomial, what do you get? When you multiply a monomial by a trinomial what do you get? And so on...**

Okay, now let's look at a situation that you'll be working with a lot in this book: multiplying a binomial by a binomial. The important thing to realize is that the Distributive Property still applies.

12. **Multiplying a binomial by a binomial is just like multiplying a two-digit number by a two-digit number. Multiply 37 by 49, using whatever method you like. Now, how many individual multiplications did you have to do to get the answer?**

The important thing about using the Distributive Property to multiply a binomial by a binomial is that you have to multiply *each term of the one polynomial by each term of the other polynomial.* That means that — just as in Problem 12 — when you multiply two binomials, you have to do four individual multiplications and then add the terms that you get from those multiplications.

The only real trick is keeping track of which of the four necessary multiplications you've done so that you don't do the same one twice. There's a classic method that every algebra student learns for keeping track of those multiplications. It's called *FOIL.* FOIL is an acronym for "First, Outside, Inside, Last." Here's how it works:

Suppose you're multiplying $(x + 5)(x + 3)$.

As Step One, you multiply the two *first* terms by each other:

$$(x + 5)(x + 3) = x^2$$

... then you multiply the two *outside* terms by each other:

$$(x + 5)(x + 3) = x^2 + 3x$$

... then you multiply the two *inside* terms by each other:

$$(x + 5)(x + 3) = x^2 + 3x + 5x$$

... then you multiply the two *last* terms by each other:

$$(x + 5)(x + 3) = x^2 + 3x + 5x + 15$$

... and finally, whenever possible (which will be fairly often), you simplify the resulting polynomial, generally by adding the two middle terms:

$$(x + 5)(x + 3) = x^2 + 3x + 5x + 15 = x^2 + 8x + 15$$

Here are a few problems to try out the FOIL method on:

13. $(x + 7)(x + 5)$

14. $(x + 12)(x + 2)$

15. $(2x + 4)(x + 9)$

I'm going to show you another method for multiplying polynomials. In my experience, most students prefer to use FOIL when they're multiplying two binomials, but this next method — I'll call it the box method — is very helpful when you need to multiply larger polynomials, so I'll go ahead and show it to you now. It's really quite simple. In order to multiply $(x + 3)$ by $(x + 5)$, you set up a box that looks like this:

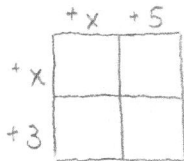

(The "+" signs are not strictly necessary. I like to use them to remind myself that all four terms are positive, because sometimes you'll be multiplying polynomials with negative terms.)

Then you do the four multiplication problems involved and put the answers in the boxes:

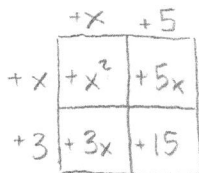

... and then you write down the final polynomial, simplifying it if appropriate: $x^2 + 8x + 15$.

Here are three to try using the box method:

17. $(x + 4)(x + 7)$

18. $(x + 3)(4x + 13)$

19. $(y + 4)(3x + 6)$ **(Be careful when it comes to simplifying this one.)**

As I say, the box method is probably a little burdensome when you're multiplying binomials, but it will come in handy soon.

Now I'll ask you to try some with negative numbers. This is really not difficult, but it will be very helpful if you do think of polynomials as having negative or positive terms rather than being strings of addition and subtraction. I'll do a quick example problem using FOIL and then you can try some.

I'll use the problem $(x - 7)(- 2x + 4)$.

$$(x-7)(-2x+4) = -2x^2 \qquad (pos. \cdot neg. = neg.)$$

$$(x-7)(-2x+4) = -2x^2 + 4x \qquad (pos. \cdot pos. = pos.)$$

$$(x-7)(-2x+4) = -2x^2 + 4x + 14x \qquad (neg. \cdot neg. = pos.)$$

$$(x-7)(-2x+4) = -2x^2 + 4x + 14x - 28 \qquad (neg. \cdot pos. = neg.)$$

$$(x-7)(-2x+4) = -2x^2 + 18x - 28$$

In case this isn't clear, I wrote it out like that just to show you the steps I go through in my head when I'm doing the problem. In my own notebook, I would not write out the problem five times, I wouldn't use all those arrows, and I certainly wouldn't write "pos. • pos. = pos." and so on. Here's what I might actually put in my notebook:

$(x - 7)(- 2x + 4) = - 2x^2 + 4x + 14x - 28 = - 2x^2 + 18x - 28$

At a minimum, your work should look like this:

$(x - 7)(- 2x + 4) = - 2x^2 + 18x - 28$

Do the following multiplication problems, remembering to simplify where appropriate:

20. $(8x - 2)(3x + 4)$

21. $(- 4x - 3)(x - 3)$

22. $(x + 11)(- x + 4)$

23. $(x - 11)(- x - 4)$

24. $(- x + 11)(x - 4)$

25. $(- x - 11)(x + 4)$

26. $(3x^2 - 6)(x + 5)$

27. $(3x^2 - 6)(x^2 + 5)$

28. $(x - 9)(x^3 + 6)$

29. $(- 8x^3 + 4)(- 2x^3 - 5)$

30. $(5x^2 - 2x)(x^2 + 3x)$

31. $(5x^2 - 2x)(x^2 + 3)$

32. $(y - 6)(x - 5)$

33. Sometimes when you multiply a binomial by a binomial your answer has three terms and sometimes it has four terms. Looking back at the problems that you just did, come up with a guideline for when the answer will have three terms and when it will have four.

Multiplying polynomials of more than two terms is not especially tricky. There is no little trick comparable to FOIL, but there are at least two options for how to approach these problems.

Suppose you were multiplying $(x - 6)$ by $(3x^2 + 2x - 8)$. One option is to set up a box (which will be a rectangle this time instead of a square) and multiply, like so:

Then you write the resulting polynomial, $(3x^3 + 2x^2 - 18x^2 - 8x - 12x + 48)$ and (of course!) simplify it to $(3x^3 - 16x^2 - 20x + 48)$.

Or, if you prefer, you can set it up like an old-school multiplication problem:

$$3x^2 + 2x - 8$$
$$x - 6$$

... and then do each of the individual multiplications (- 6 times - 8, - 6 times 2x ...), so that it ends up looking like this:

(Notice how I shifted over the second set of multiplications so that terms of like degree would line up and could be added easily.)

Do the following multiplication problems using whatever method you like. (You can invent your own way of keeping track of which multiplications you've done, if you like.)

34. $(x + 7)(4x^2 - 6x + 4)$

35. $(- x + 3)(- 2x^2 + 7x - 8)$

36. $(2x + 10)(3x^2 + 5x - 10)$

37. $(3x - 4)(5x^2 - 3x + 6)$

38. $(x^2 + 7x)(3x^2 + 8x + 9)$

Of course, your boxes don't always have to be two squares by three squares:

39. $(x + 7)(2x^3 + 4x^2 - 6x + 3)$

40. $(4x^2 + 8x - 7)(2x^2 - 6x - 3)$

41. $(5x + 3)(4x^4 + 8x^3 - 3x^2 + x - 1)$

By the way, if you're using the box method, I'm sure you've noticed this already, but it's worth remarking on the fact that — generally speaking — the terms that can be added to each other show up in diagonal rows of compartments within the box. This pattern will come in handy later on when you start factoring polynomials.

It's also possible to multiply more than two polynomials together. All you have to do is multiply the first polynomial by the second using whatever method you like, then multiply the resulting polynomial by the third, and so on.

42. Try multiplying (x)(x + 6)(2x - 4). Now, the Associative Property of Multiplication tells us that we should be able to do a string of multiplications in any order that we want, so, once you've done it in the first order, try doing that same multiplication problem in this order: (x + 6)(2x - 4)(x). Hopefully you'll get the same result!

Do the following multiplication problems:

43. x(3x - 4)(2x + 8)
(By the way, when you have a monomial in a multiplication string like that, you don't have to write parentheses around it. You can if you want to, though.)

44. (x + 2)(x - 3)(x + 5)

45. x(3x + 7)(2x^2 - 4x + 3)

46. (x + 3)^3

There are just a couple more things I'd like to cover in this lesson. The first is squaring binomials. You're just multiplying them by themselves, so at first you may want to write out, for example, (2x + 3)^2 as (2x + 3)(2x +3), but you'll find that there's a convenient little shortcut for figuring out the final, simplified trinomial.

47. (x + 7)^2 48. (x + 4)^2

49. (x - 7)^2 50. (x - 4)^2

51. (- x + 3)^2 52. (2x + 4)^2

53. (3x - 4)^2 54. (x^2 + 3x)^2

55. Just by looking at a binomial square like (x + 7)^2, how can you tell what the first term of the final trinomial will be? How can you tell about the second and third terms?

Before you go on, I'd like you to look back at the problems you just did and realize that the last term of a binomial square is always positive (because you're either multiplying a positive times a positive or a negative times a negative to get it) and that it's always a perfect square. These facts will come in handy later on when you're factoring polynomials.

56. Based on the problems you just did, do you think that it's possible for a binomial square to have anything other than three terms once it's multiplied out? Why or why not?

57. So far you've found that a binomial times a binomial can give you a trinomial or a four-term polynomial. There is one particular situation in which a binomial times a binomial can give you another binomial. See if you can figure out what binomial to multiply $(x + 3)$ by in order to give you another binomial as an answer. Your hint is that you will initially get a four-term polynomial but that the two middle terms will be opposites of one another and will therefore disappear.

This particular situation will also be important your factoring work later on, where I'll call it *the difference of squares*.

Try a few of these types of multiplication problems to see how they work:

58. $(x + 5)(x - 5)$

59. $(3x + 4)(3x - 4)$

60. $(2x + 9)(2x - 9)$

61. $(3x^2 + 5x)(3x^2 - 5x)$

62. Looking at your answers to Problems 58 through 61, why do you think I would call this type of thing "the difference of squares?"

63. Write a **Note to Self** about **multiplying polynomials**. You can explain whichever method or methods are your favorites (FOIL, boxes, old-school multiplication, or anything you came up with yourself). As with your last Note, I think that the examples will probably be the most important part.

64. Here's another problem like the final one in the last lesson, where your job is to use the algebra skills you're learning to prove that something is always true. This problem is also courtesy of Harold Jacobs, who got it from Sam Loyd — that's how these things work.

Take three consecutive numbers, such as 4, 5, and 6. If you square the middle number you get 25. If you multiply the first number by the third, you get 24 — one less than 25. Choose a few sets of three consecutive numbers in order to check that something similar seems to happen. Now you're going to use algebra to prove that it will always happen. There are several ways to go about this, but the simplest is probably to use x to stand for the middle of the three numbers...

REVIEW

1. Sketch the graph of the following equation:

$$\frac{y^2}{25} - \frac{x^2}{16} = 1$$

2. Estimate the value of $-2\sqrt{26}$.

3. Simplify the following expression:

$$\sqrt{45} - \sqrt{180}$$

4. Solve the following inequality:

$$\frac{3y - 15}{5} \geq 2y + 4$$

5. Simplify the following expression:

$$\left(\frac{15 - |2 - 10|}{3^2 - 2}\right)^2$$

6. Simplify the following expression:

$$\frac{7n}{3} + \frac{2n}{5} - \frac{5n}{2}$$

7. Solve the following equation:

$$3.5(n - 4) + 5 = 5n$$

8. What is .5% of 675?

9. Resourceful Odysseus was mixing wine and water to pour out a libation to the gods. Wine cost him 14 drachmas per liter and water cost him 3 drachmas per liter. (He had to pay the water carrier.) If he paid 76 drachmas and wound up with 7 liters of his mixture, what was the ratio of wine to water in the mixture?

10. Here's another puzzle from BrainBashers.com:

Find the hidden nine-letter word in the array below. You can move any direction, including diagonally.

H	L	E	C
L	E	H	E
A	A	N	G
E	N	A	G

(Though I found only one nine-letter word, it's possible that there are others.)

4 DIVIDING POLYNOMIALS

If you can multiply two polynomials together, then it stands to reason that you can divide one polynomial by another. In this lesson I'll show you a couple of techniques for dividing polynomials — though, once again, you may come up with a method that you like better. Just remember that dividing polynomials is fundamentally the same as dividing anything else: the question is always, "How many of this one thing go into this other thing?" Only, in this case, the answer to that question will itself be a polynomial.

I'm going to show you how to divide $(x^2 - x - 12)$ by $(x + 3)$. While I show you that, you should divide $(x^2 + 3x - 10)$ by $(x + 5)$ using the same steps.

1. **Perhaps the first question to ask is this: how many terms do you expect the answers to $(x^2 - x - 12)$ divided by $(x + 3)$ and $(x^2 + 3x - 10)$ divided by $(x + 5)$ to have? In order to make an educated guess, look back at all the multiplying of polynomials that you did in the last lesson.**

Before we start dividing, note that the problem $(x^2 - x - 12)$ divided by $(x + 3)$ is usually written one of two ways:

$$\frac{x^2 - x - 12}{x + 3} \qquad \text{or} \qquad x + 3 \overline{)x^2 - x - 12}$$

The first method that I'll show you for doing division problems involves using boxes to basically do the reverse of what you did in the last lesson.

The first step is to make your box. As I hope you realized when you did Problem 1, the answers to these problems are probably binomials (well, I *know* they're binomials because I made them up), so I'll use a two-by-two box. This is all that I will put in the box at first:

2. **Make a similar box for the problem $(x^2 + 3x - 10)$ divided by $(x + 5)$. (For Problems 3 through 7 you'll continue filling in that same box — you don't have to draw new ones. It's up to you how you want to number the problems on your page.)**

Notice that the left side of the box is blank. That's where the answer will go. (If you prefer to put $(x + 5)$ along the side of your box and the answer on the top, go ahead and do so — it doesn't make any difference.) Second, notice that the full polynomial $(x^2 - x - 12)$ is *not* written in my box — only the first term. You'll see why soon.

The next step is that I ask myself, "What times **x** will give me **x²**?" Well, **x** times **x** gives me **x²**, so I'll fill that part of the answer in, like so:

3. **Ask yourself a (very) similar question for the problem that you're doing and fill in the appropriate part of the box.**

All right, for this next step I'm going to remember that, even though I'm dividing, this box works exactly like a multiplication box (in a sense, what we're really doing here is reconstructing a multiplication problem), and for a moment I'm going to think of it as a multiplication box. The question I'm now asking myself is, "What's **3** times **x**?" Well, it's just **3x**, so I'll write that in the appropriate part of the box, like so:

4. **Fill in the corresponding part of the box that you're working on.**

The next step is probably the trickiest. In order to do it, think back to the multiplying of binomials that you did in the last lesson. Remember that there was almost always a final step where you added together two middle terms, simplifying your polynomial answer. Now I'm going to do the same thing in reverse. The polynomial that's developing inside my box so far looks like this: ($x^2 + 3x$ ...) Ultimately, it should be ($x^2 - x - 12$). So the question is, "What would I need to add to **3x** to get **- x**?" And the answer to that question is **- 4x**. I'll fill that into my box:

5. **Ask yourself a similar question and fill in the appropriate part of the box you're working on. Remember that the polynomial that you're aiming to get when you add up all four compartments is ($x^2 + 3x - 10$).**

The rest is easy. Next question: "What times **x** is **- 4x**?" Answer: **- 4**. That goes here:

6. **Fill in the corresponding part of the box for the problem that you're working on.**

Notice that we've now solved the problems we've been working on. The answer to $(x^2 - x - 12)$ divided by $(x + 3)$ is $(x - 4)$. The last couple of steps are just to check our work.

I fill in the last empty compartment of my box with the answer to the multiplication problem **(3)(- 4)**:

... and I'm very relieved to see that the answer is **- 12**, since the original dividend that I was working with was $(x^2 - x - 12)$. I run a mental check: "Does the polynomial $(x^2 + 3x - 4x - 12)$ simplify to $(x^2 - x - 12)$?" It does. In a way, you could think of **+ 3x** and **- 4x** as being the "hidden terms" in $x^2 - x - 12$. The very last thing that I do is make sure to write the original division problem with its answer somewhere near my box, so that my final work looks like this:

$$\frac{x^2 - x - 12}{x + 3} = x - 4$$

7. **Fill in the final compartment of the box you're working, mentally check that your dividend simplifies to $(x^2 + 3x - 10)$, and write the division problem and answer near your box.**

I know that seemed like a pretty long process, but we were going step by step — it will go a lot faster once you've practiced it a couple of times.

Do the following division problems using the box method:

8. $(x^2 - 2x - 15)$ divided by $(x + 3)$

9. $(2x^2 + 7x + 6)$ divided by $(x + 2)$

10. $(-8x^2 - 10x + 3)$ divided by $(2x + 3)$

11. $(15x^2 - 14x + 3)$ divided by $(3x - 1)$

12. $(3x^2 + x - 2)$ divided by $(-3x + 2)$

13. $(3x^4 + 7x^3 + 2x^2)$ divided by $(x^2 + 2x)$
(This may look a little bit more complicated, but it can actually be solved using a two-by-two box just like the ones you made in Problems 8 – 12.)

14. $(2x^3 - 4x^2 + 6x - 12)$ divided by $(x - 2)$
(Box-method division is actually *easier* to do when your dividend is a four-term polynomial.)

15. $(2xy + 6x - 2y - 6)$ divided by $(y + 3)$

16. $(x^2 - 25)$ divided by $(x + 5)$
(Set this one up just the way you have been. The question here is, "What do the hidden terms add up to?")

17. $(2x^3 + 9x^2 + 19x + 15)$ divided by $(2x + 3)$
(This one is a little harder. The answer is going to be a *trinomial*, so your box should be two compartments by three. The first thing to do will be to fill in the 2x + 3 and then just the $2x^3$ in the upper left compartment. Then follow the same steps you've been using for the two-by-two boxes. There will be four hidden terms this time instead of two.)

18. $(12x^3 - 19x^2 - 11x + 20)$ divided by $(-3x^2 + x + 4)$
(This time the answer will be a binomial.)

19. $(4x^4 + 14x^3 + 20x^2 + 15x + 3)$ divided by $(2x^2 + 3x + 3)$
(We're getting trickier now. Start the same way you have been, but know that this time the answer will be a trinomial. There will be two hidden terms that add up to $14x^3$, *three* that add up to $20x^2$, and two that add up to 15x.)

20. $(3x^4 - 17x^2 + 26x - 12)$ divided by $(3x^2 - 6x + 4)$
(This is the trickiest one of the lot. The answer will be a trinomial.)

Excellent work.

I told you how many terms your answer was going to have for the problems in that last set, but you don't really need that information in order to use the box method. You can expand the box as necessary as you go along.

For instance, your work for Problem 20 could start out looking like this:

$$\begin{array}{|c|c|c|} \hline +3x^2 & -6x & +4 \\ \hline +3x^4 & & \\ \hline & & \\ \hline \end{array}$$

... and then you could figure out how many rows you would need as you worked through it.

Do the following division problems. I'm not going to tell you how many terms the answers will have.

21. $(3x^3 - 14x^2 + 12x - 16)$ divided by $(x - 4)$

22. $(5x^4 - 19x^3 + 15x^2 - 13x + 12)$ divided by $(x^2 - 4x + 3)$

23. $(- 3x^3 + 12x + 9)$ divided by $(3x + 3)$

24. $(8x^4 - 8x^3 + 7x - 2)$ divided by $(2x^2 - 3x + 2)$

There is another method for dividing polynomials that I'd like to show you. In fact, it works more or less exactly like the long division that you learned to use years ago — I suppose I should call it old-school division.

Here's a quick reminder of some division vocabulary that I've already used in this lesson:

$$13\overline{)65} \qquad \text{quotient} = 5$$

divisor — 13   dividend — 65

25. **Just so you have the method for long division fresh in your mind so you can use it with polynomials, use long division to divide 12,949 by 23.**

You may find that you prefer the box method every time or the old-school long division method every time or that you like to mix it up. In my opinion, the long division method is especially useful when you're dealing with a dividend or a divisor that has a lot of terms.

Just as I did with the box method, I'm going to walk you through the steps of an old-school division problem with polynomials and I'll ask you to solve a similar problem along with me. I'll solve the problem $(2x^3 - 10x^2 - 36x + 56)$ divided by $(x - 7)$ and I'll ask you to solve $(4x^3 - 19x^2 - 36x + 36)$ divided by $(x - 6)$.

The first step is, naturally, to set up the problem:

$$x-7 \overline{\smash{\big)}\, 2x^3 - 10x^2 - 36x + 56}$$

26. **Set up your problem in a similar fashion. (Just as in Problems 2 – 7, Problems 26 – 32 will appear as a single division problem on your page.)**

Now I ask, "What times **x** will give me **2x³**?" Notice that I'm only concerning myself with the first terms of the two polynomials. My answer is **2x²**, and I'll fill it in like this:

$$\begin{array}{r} 2x^2 \\ x-7 \overline{\smash{\big)}\, 2x^3 - 10x^2 - 36x + 56} \end{array}$$

(Notice that I put my "**2x²**" above the "**10x²**" in the dividend. That's just like lining up place values in old-school long division.)

27. **Ask yourself the corresponding question for your problem and fill in the answer.**

Now I'm going to multiply **2x²** by *both terms* of (**x - 7**), which gives me (**2x³ - 14x²**). I'll fill in that polynomial under the first two terms of the dividend (still lining up "place values"), like so:

$$\begin{array}{r} 2x^2 \\ x-7 \overline{\smash{\big)}\, 2x^3 - 10x^2 - 36x + 56} \\ 2x^3 - 14x^2 \end{array}$$

28. **Take the same step in your problem.**

Now, just as I would if I were doing regular old long division, I'm going to subtract (**2x³ - 14x²**) from (**2x³ - 10x²**). I'm going to be careful, though, because I've studied how to subtract polynomials: **2x³** minus **2x³** is zero, but (**- 10x²**) minus (**- 14x²**) means subtracting a negative, and the answer is (**+ 4x²**). Make 100% sure you understand my reasoning, because forgetting how to subtract negatives is the easiest mistake to make in these long division problems. I'll fill in my work like this:

$$\begin{array}{r} 2x^2 \\ x-7 \overline{\smash{\big)}\, 2x^3 - 10x^2 - 36x + 56} \\ 2x^3 - 14x^2 \\ \hline 4x^2 \end{array}$$

29. **Take the corresponding step in your problem, being careful of the same thing.**

Now I'll "bring down" the next term of the dividend just as I would in regular long division. (Actually, I object to the whole notion of "bringing down" in doing long division. I think it obscures what's really happening, but since most math students learn to use that method, I'm going to stick with it here.) My problem now looks like this:

$$
\begin{array}{r}
2x^2 \\
x-7 \overline{\smash{\big)}\, 2x^3 - 10x^2 - 36x + 56} \\
\underline{2x^3 - 14x^2 } \\
4x^2 - 36x
\end{array}
$$

30. **Take the corresponding step in your problem.**

I bet you can predict what I'll do next, since I'm really repeating the steps I've already done. So I am going to take several steps at once. I ask myself, "What times **x** will give me **4x²**?" The answer to that is **4x**, so I fill that in as part of the quotient, then multiply **4x** by both terms of (**x - 7**), which gives me (**4x² - 28x**). I subtract (**4x² - 28x**) from (**4x² - 36x**), which gives me (**- 8x**). I fill that in and bring down the last term of the dividend. All of those steps look like this:

$$
\begin{array}{r}
2x^2 + 4x \\
x-7 \overline{\smash{\big)}\, 2x^3 - 10x^2 - 36x + 56} \\
\underline{2x^3 - 14x^2 } \\
4x^2 - 36x \\
\underline{4x^2 - 28x } \\
- 8x + 56
\end{array}
$$

31. **Take the corresponding steps in your problem.**

The final steps follow the same pattern. The only thing to be careful of here is that the answer to "What times **x** will give me **- 8x**?" is itself a negative number, so I need to make sure that I write it as negative in the quotient. Fortunately, (**- 8**) times (**x - 7**) is precisely (**- 8x + 56**), so the last subtraction gives me a result of zero. My completed work looks like this:

$$
\begin{array}{r}
2x^2 + 4x - 8 \\
x-7 \overline{\smash{\big)}\, 2x^3 - 10x^2 - 36x + 56} \\
\underline{2x^3 - 14x^2 } \\
4x^2 - 36x \\
\underline{4x^2 - 28x } \\
- 8x + 56 \\
\underline{- 8x + 56} \\
0
\end{array}
$$

32. Finish the problem you've been working on.

The fact that both of our final subtractions gave us zero as a result brings me to one key difference between old-school long division and long division using polynomials: in the case of polynomials, the concept of a "remainder" is essentially useless. Remainders are one way of dealing with the fact that not all numbers are evenly divisible by all other numbers. For instance, the quotient of **13** divided by **4** can be usefully expressed as **3** with a remainder of **1**. A better way of expressing that same idea is probably that the quotient of **13** divided by **4** is **3¼** or **3.25**. If you really wanted to, I suppose you could express the remainder of a polynomial division problem as a fraction using polynomials, but I'm not going to ask you to explore that concept in this book. The fact that we're not going to get into polynomial division with remainders has a handy consequence for you: you don't have to check your polynomial division problems. Since none of the problems in this book will have remainders, if you've done your work correctly, you won't get a remainder. (I suppose there's always the possibility that you worked incorrectly in such a way as to come up with an incorrect answer that still has no remainder, but that's very unlikely to happen.)

33. Before I give you some problems to practice long division on, use the box method to solve the same problem you just solved using long division: $(4x^3 - 19x^2 - 36x + 36)$ divided by $(x - 6)$.

34. Now compare the work that you did solving that problem using the box method with the work you did solving it by the long division method. What do you notice about the hidden terms from the box method when you study the long division method?

Solve the following problems using long division:

35. $(21x^3 + 2x^2 + 10x + 12)$ divided by $(3x + 2)$

36. $(-24x^3 + 24x^2 - 54x + 24)$ divided by $(-6x + 3)$

37. $(9x^3 - 18x^2 + 18x - 9)$ divided by $(3x^2 - 3x + 3)$
(In this case, the divisor has three terms. This actually makes the problem easier than the previous ones. Each time you multiply, you should end up with a trinomial instead of a binomial.)

38. $(15x^3 - 9x - 6)$ divided by $(3x - 3)$
(I strongly recommend filling in the "missing" x^2 place in the dividend with $0x^2$ so that your initial problem looks like this:

$$3x - 3 \overline{\smash{)}\, 15x^3 + 0x^2 - 9x - 6}$$

In fact, the problem would be hard to do without that $0x^2$.)

39. $(6x^3 - 16x - 16)$ divided by $(3x^2 + 6x + 4)$ (Again, fill in that $0x^2$.)

40. It probably won't surprise you to learn that I made up these division problems by doing multiplication problems. It was a little more challenging for me to make up the last two that had "missing" x^2 terms in the dividends. Try making one up yourself. The divisor should be a binomial, the quotient should end up being a trinomial, and the dividend should be a trinomial as well — with a "missing" term. It was much easier for me to think about how this worked when I used the box method for the multiplication. Once you've finished your problem, give it to a classmate to solve.

Solve these problems using long division:

41. $(2x^4 + 5x^2 - 14x + 15)$ divided by $(x^2 + 2x + 5)$
(As before, you'll need to fill in the "missing" terms of the dividend.)

42. $(2x^4 + 32x - 24)$ divided by $(- x^2 + 2x - 6)$
(Remember those "missing" terms. By the way, I keep using the word "missing" in quotes like that because I don't believe they're really missing, but it's a useful way of thinking about it.)

43. Use multiplication to create a problem like the one I just gave you with two "missing" terms in the dividend.

Do one more using long division:

44. $(- 6x^5 + 8x^4 - 5x^3 - 11x^2 + 19x - 5)$ divided by $(- 3x^2 + 4x - 1)$
(As you'll probably figure out pretty quickly, this time it will make sense to "bring down" two terms instead of one the first time you "bring down" — either that or put a $0x^2$ in the quotient.)

For these last two problems, which involve more than one variable, either long division or the box method will work, but you have to be a little creative about how you apply either one:

45. $(2x^2y - 3x^2 + 8xy - 12x + 6y - 9)$ divided by $(2y - 3)$

46. $(12x^2 + 5xy - 2y^2)$ divided by $(3x + 2y)$

47. Write a **Note to Self** about **dividing polynomials**. Describe whichever methods you like to use, including any you've come up with. Clearly explained examples will be key. If you choose to explain the long division method, be sure to pick an example that involves putting in a "missing" term.

REVIEW

Name the coefficient and degree of the following monomials:

1. $3x^5$

2. $-7x$

Simplify each of the following polynomials and name its degree:

3. $x - x^5 + 3x^2 - 5x^5 + 10x - 9$

4. $4x^8 - 2x^6 - 2x^8 + x^3 + 15 + 2x^6 - 2x^8$

5. Add $(4x^5 + 3x^4 + 2x^3 + x - 1)$ to $(-3x^5 - 5x^4 + 6x^3 + 10x - 15)$.

6. Subtract $(2x^4 + x^3 - 3x^2 + 7x - 5)$ from $(5x^4 + x^3 - 2x^2 + x + 9)$.

7. Subtract $(4x^8 + 3x^6 - 2x^5 + x^3 - 3x^2)$ from $(5x^6 + 3x^5 - x^4 + x^2 - 3)$.

8. Multiply $(x - 3)$ by $(4x + 5)$.

9. Multiply $(3x - 6)$ by $(2x^2 - 7x + 3)$.

10. Multiply $(x + 3)$ by $(x - 5)$ by $(2x + 1)$.

11. Multiply out $(3x + 2)^2$.

12. Divide $(6x^3 - 11x^2 + 13x - 15)$ by $(2x - 3)$.

13. Divide $(2x^4 - 13x^2 + 21x - 3)$ by $(x^2 - 3x + 3)$.

3

FACTORING POLYNOMIALS & SOLVING QUADRATIC EQUATIONS

Some of what you'll be learning in this lesson is actually a review of stuff that you learned back in *Jousting Armadillos* — only at that point you had no idea that these skills had anything to do with polynomials. What you're going to be doing is looking for common factors among the terms of polynomials. Of course, each of the terms of a polynomial is itself a monomial and, in fact, you have done quite a lot of factoring of monomials, though you didn't know at the time that they were monomials.

I'm talking about problems like this one, which is taken straight out of *Jousting Armadillos*, Chapter 4, Lesson 2:

"Find the greatest common factor of $16n^3$, $28n^2$, and $32n^5$."

In order to do that problem, you learned to break down each of those monomials (that's what they are, right?) into their prime factors, like so:

$16n^3 = 2^4 n^3$
$28n^2 = 2^2 \cdot 7n^2$
$32n^5 = 2^5 n^5$

1. **What is the greatest common factor of those three monomials?**

Now, for the work that you'll be doing in this chapter, it's not so important that you should prime factorize monomials as that you should be able to list what all of their factors are. Ideally, you'll get to the point where you can glance at most monomials and be able to list their factors in your head.

Let's look, for example, at $16n^3$. A list of its factors would look like this:

$8n$
n^2
4
16
$4n^2$
$2n^2$
$8n^2$
$16n$
1
n^3
$16n^3$
$16n^2$
n
$8n^3$
2
$2n$
$4n^3$
$4n$

2. I have listed those factors randomly. In fact, they really belong in pairs. Rewrite that list in pairs that, when you multiply them together, give you $16n^3$. (That's what makes them factors of $16n^3$, right?)

(As a quick aside, I've really only listed half of the factors of **$16n^3$**. For instance, (**- 4n**) and (**- $4n^2$**) are a factor pair of **$16n^3$**. When you list all of the factors of a positive monomial, you don't have to list the negative pairs, but keep in mind that they exist.)

Notice that one of the pairs you made in Problem 2 is really pretty pointless. It's true that **$16n^3$** and **1** are both factors of **$16n^3$**; every number or monomial (or polynomial) has itself and **1** as factors. Let's just assume that you know that's true and, from now on, when I ask you to list the factors of a monomial, you can leave **1** and the monomial itself off of the list.

List the factors of the following monomials. I strongly suggest listing them as pairs — I think you'll find that makes it easier to keep track of whether you've found all the factors.

3. x^3

4. x^5

5. 24

6. 15x

7. $4x^3$

8. $49x^2$

9. x^3y

10. xy^3

11. 100x

12. One of the monomials that you just factored had an unpaired factor (or, if you prefer, a factor that was paired with itself). Which monomial was it? What special category does a monomial have to fall into in order to have an odd number of factors (or a factor paired with itself, if you prefer to think of it that way)?

Okay, now it's time for you to begin factoring polynomials. Consider, for example, the polynomial (**6x + 2**).

13. You can clearly think of (6x + 2) as two monomials added to one another. Those monomials have a common factor (other than 1) — what is it?

The fact that **6x** and **2** have **2** as a common factor means you can rewrite the polynomial (**6x + 2**) like so: **2(3x + 1)**.

I just said that you were going to begin factoring polynomials, but of course I lied. Because you have been doing this kind of simplifying:

(6x + 2) = 2(3x + 1)

... since *Jousting Armadillos*. It's just that up to this point you didn't realize you were factoring polynomials. Up until now, I've called this kind of thing "undistributing." I still like that name and I may still use it sometimes, but the proper name for it is *factoring out the greatest common factor of a polynomial.* (Kind of a clunky name, huh? Maybe you can see why I prefer "undistributing.") When people do this kind of simplification:

$(6x + 2) = 2(3x + 1)$

... they'll say, "I factored a **2** out of (**6x + 2**)."

I want to point out a couple of things before I give you some polynomials to practice on. First of all, the monomial that you factor out of your polynomial may include a variable. For example, if I were to factor (**8x² - 2x**), I would do it like so:

$(8x^2 - 2x) = 2x(4x - 1)$

(Be sure to check mentally by multiplying that I've factored that polynomial correctly.)

Second, you may sometimes have more than one option for factoring. For instance, (**8x² - 4x**) could be factored in any of the following ways:

$(8x^2 - 4x) = 2(4x^2 - 2x)$
$(8x^2 - 4x) = 2x(4x - 2)$
$(8x^2 - 4x) = 4(2x^2 - x)$
$(8x^2 - 4x) = 4x(2x - 1)$
$(8x^2 - 4x) = -2(-4x^2 + 2x)$
$(8x^2 - 4x) = -2x(-4x + 2)$
$(8x^2 - 4x) = -4(-2x^2 + x)$
$(8x^2 - 4x) = -4x(-2x + 1)$

(Again, run a quick mental check to see that my factoring is correct.) In general, you should make a habit of choosing to factor out the greatest common factor. There maybe certain situations where factoring out the greatest common negative factor is useful, but generally you should choose the greatest common positive factor. So the best choice for (**8x² - 4x**) is (**4x**)(**2x - 1**). Unless you're specifically asked to do something else, or you feel like the problem-solving situation demands something else, assume that you should factor out the greatest common positive factor.

Factor out the greatest common factor of the following polynomials:

14. $(15x - 5)$

15. $(-18x + 6)$

16. $(32x^2 + 24x)$

17. $(-x^2 + 3x)$

18. $(9x^2 + 12y^2)$

19. $(8x^3 - 8x^2)$

20. $(7xy^2 + 13xy)$

21. $(6x^2y - 4xy^2)$

22. $(8x^2 + 12x - 2)$ **(Yes, it's definitely possible to factor a monomial out of a trinomial. I don't think you need a specific set of instructions on how to do it. I'll just say that you should end up with a monomial times a trinomial.)**

23. $(-24x^3 - 6x^2 + 32x)$ 24. $(9x^9 + 6x^6 - 3x^3)$

25. $(6x^3y + 10x^2y - 14xy)$ 26. $(10x^{10} + 6x^6 - 4x^4 + 2x^2)$

27. $(25x^5 - 100x^4 + 200x^3 - 75x^2 + 50x)$ 28. $(4x^2 - 6x + 11)$

I put in Problem 28 to make the point that it's not always the case that the terms of a polynomial *have* a common factor (other than **1**, which is, of course, a factor of any monomial). In the coming lessons, you'll learn ways to factor polynomials whose terms don't have common factors. But sometimes — and this is the case with ($4x^2 - 6x + 11$) — polynomials don't have any factors other than themselves and **1**. Those polynomials are, not surprisingly, called *prime polynomials*.

I know I told you that you should generally factor the greatest *positive* factor out of a polynomial, but there are certain situations where it is useful to factor **-1** out of a polynomial. This can be done even when a polynomial is prime.

Factor -1 out of the following polynomials:

29. $(2x + 5)$ 30. $(y + x)$

31. $(x - y)$ 32. $(3x^2 - 7x + 8)$

33. **What effect does factoring -1 out of a polynomial have on the signs of the terms of the polynomial?**

Sometimes, even when a polynomial is prime, it can be handy to find a common factor of some of the terms. In any case, I'd like you to know that this is a possibility. For example, consider the polynomial ($6x^2 + 3x - 7$). Now, this is a prime polynomial, but notice that $6x^2$ and $3x$ have a common factor. This means it's perfectly permissible to rewrite that polynomial in the following way:

$6x^2 + 3x - 7 = 3x(2x + 1) - 7$

34. **Double-check my work. Distribute the 3x in the expression:**

$3x(2x + 1) - 7$

... and see what you get.

Now, rewriting ($6x^2 + 3x - 7$) as ($3x(2x + 1) - 7$) is not, strictly speaking, the same thing as factoring ($6x^2 + 3x - 7$) because I haven't found a factor pair. I guess I would call it something like "partial factoring." (As far as I know it doesn't have an official name.) Don't get the impression that you should go around trying to partial factor every polynomial you encounter, but do keep in mind that it's possible because it can be handy.

Here are a few more of examples of partial factoring:

$12x^2 - 3x + 5 = 3x(4x - 1) + 5$
$7x^2 + 56x + 2 = 7x(x + 8) + 2$
$5x^2 + 6x + 9 = 5x^2 + 3(2x + 3)$

Partial factor the following polynomials:

35. $5x^2 + 25x + 14$ 36. $4x^2 - 12x + 5$

37. $5x^2y + 3xy + 7$ 38. $x^2 + 7x + 3$

39. $x^2 + 10x + 5$ (In this case, there are two possible ways to partial factor
the expression; see if you can find them both.)

40. It's time for a **Note to Self** about **common factors in polynomials**. It should explain, using examples, how to factor out the greatest common factor of the terms of a polynomial. It doesn't need to explain how to do partial factoring.

These last two problems both ask you to use your factoring skills to prove things in the way that you did at the end of two lessons in Chapter 2. They're both potentially challenging (especially Problem 42), so you might think about tackling them with partners. In both cases I've broken the problems up into lettered paragraphs with the idea that you should do them a paragraph at a time.

41. a) Get a calculator. Enter a random six-digit number (but write the number down — you'll have to enter it again). Divide that number by 7. Does 7 go into your number evenly? Try 13. Does 13 go in evenly? How about 11? Try the same thing for two other random six-digit numbers. Do 7, 11, and 13 go into them evenly?

    b) Hopefully you saw that most randomly chosen six-digit numbers tend not to be divisible by 7, 11, and 13. Now choose a six-digit number where the digits repeat as a set of three, such as 546,546. Divide your number by 7, 11, and 13. Do they go in evenly? Try two more of those numbers. Do 7, 11, and 13 go into them evenly?

    c) I claim that 546,546 can be represented by the expression 1,001x, with x equal to 546. In fact, any similar number can be represented by the expression 1,001x, with x equal to the three-digit number in the last three places. Explain why this is true.

    d) Use the fact that any such number can be represented by the expression 1,001x, along with your factoring skills, to prove that 17, 11, and 13 will always go evenly into a number like 546,546.

42. a) If a number that has 5 as its last digit is squared, the resulting number always ends in 25:
$$15^2 = 225$$
$$25^2 = 625$$
$$35^2 = 1,225$$
$$45^2 = 2,025$$
$$55^2 = 3,025$$
$$65^2 = 4,225$$
$$75^2 = 5,625$$
$$85^2 = 7,225$$
$$95^2 = 9,025$$

There is also a trick for predicting the other digits of the squared form from the first digit of the unsquared form. (In other words, there's a consistent pattern that allows you to get 6 from 2, 12 from 3, 20 from 4, 30 from 5, and so on.) Study the list on the previous page and explain what that trick is.

b) I believe that any number the last digit of which is 5 can be expressed as (10x + 5), where x is equal to the other digits of the number. (For example, in 65, x would be equal to 6; in 105, x would be equal to 10.) Explain why this is true.

c) Using the fact that any number ending in 5 can be expressed as (10x + 5), prove that the trick you found in Part A will work for any number ending in 5. Start by taking the expression (10x + 5) and squaring it, since that's what's happening to each of the numbers on the left-hand side of the pattern in Part A. Then take the resulting expression and use the partial factoring skills that you practiced in Problems 35 through 39. Explain how this proves that the trick you found in Part A will always work.

REVIEW

Tell whether the graph of each of the following equations would be a straight line, a parabola, a hyperbola, an exponential curve, or a non-functional curve. Assume that x is the independent variable, y is the dependent variable, and any other letters are constants.

1. $.7x^2 - 6 = (y + 2)^2$

2. $y + 3y = \dfrac{7}{x + 5}$

3. $y - m = n^x$

4. $(x - a)^2 = \dfrac{y}{b}$

Apply the appropriate Laws of Exponents to the following expressions:

5. $7m^3n^2 \cdot 5m^{-2}n^{-3}$

6. $(3x^2y^{-4})^3$

7. Rewrite this expression so that it has neither negative exponents nor parentheses:

$$\left(\dfrac{2a^{-2}b^3}{c^{-3}d^2}\right)^{-3}$$

8. Graph the equation $y = -2(x + 1)^2 + 10$.

9. A highwayman's saddlebags are full of silver, gold, copper, and bronze coins. The ratio of gold to silver coins is 3 : 1 and the ratio of bronze to copper coins is 4 : 3. There are three more copper coins than silver coins and the total number of coins is 83. How many of each kind of coin does he have?

10. Here's a puzzle from brainden.com:

What five-digit number has the following property?

When you put the numeral 1 at the end of this number (to make a six-digit number), you get a number exactly three times as great as when you put the numeral 1 in front of the number.

2 FACTORING SECOND-DEGREE POLYNOMIALS

All right. So, like numbers, some polynomials are prime and some polynomials can be factored. (By the way, a number that isn't prime is called *composite*, although I don't know whether the same word is used for polynomials.) In the last lesson, you worked on factoring common factors out of the terms of a polynomial. So, if you looked at the polynomial $3x^2 + 12x - 9$, you'd see that you could factor **3** out of all of the terms and you'd know that it wasn't prime. At this point, you might be tempted to think that the polynomial $x^2 + 4x - 5$ is prime since the terms don't share a common factor. However, $x^2 + 4x - 5$ actually *is* factorable, and so are many similar polynomials. In this lesson, you'll learn how to factor polynomials like $x^2 + 4x - 5$.

To get started, do the following multiplication problems:

1. $(x + 5)(x + 3)$

2. $(x - 2)(x + 4)$

3. $(x - 7)(x + 2)$

4. $(x - 6)(x - 3)$

Now, take a look at your answers: hopefully they're the polynomials $(x^2 + 8x + 15)$, $(x^2 + 2x - 8)$, $(x^2 - 5x - 14)$, and $(x^2 - 9x + 18)$. Notice that none of those polynomials can be factored using the method that you learned in the last lesson — in no case do all three terms have a common factor. And yet, you know for sure that those four polynomials are not prime.

5. **How do you know that those four polynomials are not prime?**

If I were to ask you to factor $(x^2 + 8x + 15)$, what you would essentially be doing is Problem 1 in reverse. You would be "unmultiplying" $(x^2 + 8x + 15)$, in exactly the same way that, when you factor **10**, you "unmultiply" it into **2 · 5**. If you look at it another way, factoring is doing a division problem where you're not told what the divisor is — you have to find both the divisor and the quotient. So how do you go about doing that?

Do the following multiplication problems:

6. $(x + 4)(x + 3)$

7. $(x + 10)(x + 2)$

8. $(x + 3)(x + 9)$

9. **Compare each of the trinomials that you got as answers for Problems 6 – 8 with the two binomials you were originally multiplying. There is a consistent relationship between the coefficients of the second and third terms of the trinomials and the second terms of the two binomials. (In other words, in the case of Problem 6, there's a simple way to get 7 and 12 from 3 and 4 — or rather, two simple ways.) What is that relationship?**

Do these multiplication problems:

10. $(x - 6)(x - 4)$

11. $(x - 7)(x - 2)$

12. $(x - 5)(x - 4)$

13. Compare the trinomials to the original pairs of binomials for Problems 10 – 12. Does the relationship that you found in Problem 9 still hold true? If not, is there a different sort of pattern to be found?

Do these multiplication problems:

14. $(x + 5)(x - 3)$

15. $(x - 6)(x + 9)$

16. $(x - 11)(x + 3)$

17. $(x + 5)(x - 10)$

18. Once again, compare the trinomials to the binomials. Does the relationship that you found in Problem 9 still hold true? If not, is there a different pattern?

19. Use the relationship (or pattern or trick or whatever you want to call it) that you've discovered to factor $(x^2 + 13x + 36)$ into two binomials.

As you've seen, to factor a second-degree polynomial like the ones you created through multiplication in the previous problems, what you have to do is find two numbers the sum of which is the coefficient of the polynomial's second term and the product of which is the coefficient of its third term. Or, to put it in concrete terms, if you want to factor $(x^2 + 8x + 15)$, what you're looking for is two numbers whose sum is **8** and product is **15**. Those numbers are **3** and **5**. So, $(x^2 + 8x + 15) = (x + 3)(x + 5)$.

If no such two numbers exist, you can call the polynomial prime. Look at $(x^2 + 6x + 13)$: you'll find that the only factors of **13** are **13** and **1**... well, when you add **13** and **1** you definitely don't get **6**, so $(x^2 + 6x + 13)$ is prime.

I'll give you a bunch of polynomials to practice factoring in a moment, but first I want you to notice one more thing.

20. Look back at Problems 6 – 8. In those problems, the constants of the trinomials that you got as answers were all positive. So were the middle terms of the trinomials. In each case, what were the signs of the constants of the two binomials?

21. Look back at Problems 10 – 12. In those problems, the trinomials' constants were positive, but the middle terms were negative. In each case, what were the signs of the constants of the two binomials?

22. Look back at Problems 14 – 17. In those cases, the constants of the trinomials were all negative and their middle terms were sometimes positive and sometimes negative. In each case, what were the signs of the constants of the two binomials?

23. What you discovered in Problems 20 – 22 will come in quite handy in a lot of the factoring work you'll be doing. Write it down as a set of rules. (If you consider it for a minute, I hope you'll see that they really are rules because of what always happens when you multiply and add positive and negative numbers.)

Factor the following trinomials into two binomials. If a trinomial is prime, say so.

24. $x^2 - 4x - 21$ 25. $x^2 + 12x + 32$

26. $x^2 + 2x - 15$ 27. $x^2 + 3x - 15$

28. $x^2 - x - 2$ 29. $x^2 + 10x + 24$

30. $x^2 - 14x + 24$ 31. $x^2 - 5x - 24$

32. $x^2 + 12x + 36$ 33. $x^2 + 19x + 90$

34. $x^2 - 5x + 32$ 35. $x^2 - 12x + 35$

36. $x^2 + 14x + 13$ 37. $x^2 - 10x + 25$

38. $x^2 + 11x - 26$ 39. $x^2 + 2x + 1$

40. $x^2 - 12x + 32$

41. Problems 32, 37, and 39 have something in common when they're factored. What is it? What do you notice about the last terms of those three trinomials?

If you'll recall from the last chapter, when you multiply two binomials, you very often get a trinomial as a result — but not always. In one particular kind of case, you get a binomial.

42. Even though ($x^2 - 36$) is a binomial, it can be factored into two binomials very much like the trinomials you've worked with so far. You can figure out how to do this on your own. It might help if you think of ($x^2 - 36$) as having a "hidden" middle term and ask yourself what the coefficient of that term is — then factor it exactly as you did in Problems 24 – 40.

As I mentioned in the last chapter, the special situation that you just worked with in Problem 42 is called *the difference of squares* because it only applies when the binomial you're trying to factor consists of two perfect squares, such as x^2 and **36**, and one is being subtracted from the other.

Factor these polynomials:

43. $x^2 - 25$

44. $x^2 - 100$

The difference of squares is easy to factor into two binomials. The *sum* of squares, on the other hand, is impossible to factor into two binomials.

45. **Explain why $(x^2 - 16)$ can be factored into two binomials while $(x^2 + 16)$ can't.**

The method that you've learned in this lesson can be used to factor somewhat more complex polynomials, including ones with multiple variables and ones that are higher than second degree. I'm not going to give you a bunch of explicit instructions on how to do these, but instead let you try them out. For all of the problems in the next set, you're still going to be using the basic method that you've practiced so far: looking for two numbers that have a specific product and a specific sum and using those two numbers to create two binomial factors. Since these problems are somewhat more complicated, I won't give you any that are prime; they can all be factored into two binomials.

Factor the following polynomials and test to see whether your factoring is correct by multiplying the two binomials you come up with — you can do this multiplying in your head if you like.

46. $x^2 + 9xy + 20y^2$ 47. $x^2 + 7xy - 30y^2$

48. $x^4 + 3x^2 - 18$ 49. $x^4 - 3x^2 + 2$

50. $x^{10} + 10x^5 + 25$ 51. $x^6 - 4$

52. $xy + 3x + 6y + 18$ 53. $xy - 5x - 5y + 25$

Leonard Euler was probably one of the greatest mathematicians of all time. He lived from 1707 to 1783 in Europe, mostly in Russia and Prussia (which is part of modern-day Germany). He did work in all sorts of mathematical fields and he, like many other mathematicians, was very interested in prime numbers. Euler wanted to find a formula that would always give you a prime number. He proposed several such formulas, one of them being $(y = x^2 + x + 17)$. The idea is that, for any whole number x, y should be prime.

54. **Explain why the formula $(y = x^2 + 5x + 6)$ would *never* give you a prime number. (Hint: There's a reason why I put this problem in this particular lesson.)**

55. **Explain why Euler might have hypothesized that $(y = x^2 + x + 17)$ would always give you a prime number.**

As it happens, $(y = x^2 + x + 17)$ does *not* always give you a prime number, although it will go on giving you prime numbers for quite a while as you plug in larger and larger values for x. In fact, no one has yet discovered a formula that will *always* give you a prime number.

56. It's time for a *Note to Self* that I would suggest calling something like *Factoring Second Degree Polynomials, Part 1: Polynomials With First-Term Coefficients of 1*. (I know it's a pretty clunky title. You don't have to use it if you don't want to). As you can probably guess from this clunky title, in a couple of lessons this factoring work will be getting more complicated. Looking back at this lesson, you may notice that the coefficients of the first terms of the trinomials you've factored have always been 1. (In other words, I haven't asked you to try factoring anything like $2x^2 + 6x - 10$.) This Note should explain how to do what you've learned in this lesson, including the rules you wrote in Problem 23 and the difference of squares from Problems 42 – 44. You don't need to include the more complicated stuff from Problems 46 – 53.

57. Here is a very ancient problem. Solve it without guessing and checking — the method I used was to combine what I know about factoring the difference of squares with what I know about solving simultaneous equations by substitution.

Two numbers' sum is 20. The difference of their squares is 80.
What are the two numbers?

There's one last thing I'd like to point out before you move on. When we list the factors of a positive number, we generally only list its positive factors. For instance, we usually say that the factors of **7** are **7** and **1**. But in fact, the pair **-7** and **-1** are also factors of **7**. Something very similar is true of all of the polynomials that you factored in this lesson. Take, for example, the polynomial from Problem 35: $(x^2 - 12x + 35)$. When you factored it, you got $(x - 5)(x - 7)$.

58. Multiply $(-x + 5)(-x + 7)$.

So $(x^2 - 12x + 35)$ actually has two sets of factors: $(x - 5)(x - 7)$ and $(-x + 5)(-x + 7)$.

59. In order to see how this is basically the same situation as 7 actually having four factors, let x be equal to 2 in the expressions $(x - 5)(x - 7)$ and $(-x + 5)(-x + 7)$ and do the calculations. What happens in each case?

For the purposes of this book, you can continue to find only the "positive" factors of polynomials like the ones you've been working with. (*Positive* is in quotes the last sentence because the polynomials $(x - 5)$ and $(x - 7)$ don't actually have signs in and of themselves — after all, when you plugged **2** in, those expressions actually gave you negative numbers. Nonetheless, I'll call them "positive" because the **x**'s are positive. But I'll leave it in quotes.)

Okay, now for a bit of review and then it's on to quadratic equations.

REVIEW

Solve the following equations and inequalities for y:

1. $-2y - 3x = 3y + x^2 - 2$

2. $\dfrac{5(y+1)}{2} = 3x + 2$

3. $\dfrac{y}{-3} - 3x \leq -2x - 3$

4. $xy + 2y = 9$

5. Express $\sqrt{28x^2y^3}$ in simplest radical form.

6. Graph the equation $y = -\dfrac{12}{x}$.

7. Simplify the following expression:

   $\dfrac{1.2 \cdot 10^7}{3.0 \cdot 10^9}$

8. Apply the Distributive Rule to the following expression, and be sure to write your answer in simplest radical form:

   $\sqrt{3x^3}\left(\sqrt{3} + \sqrt{7x}\right)$

9. Elizabeth and Erin were having a veggiedog-eating contest. They each ate for ten minutes, but Erin ate three times as fast as Elizabeth. When the buzzer rang, they'd consumed a remarkable combined total of 600 veggiedogs. How many did each madwoman eat?

10. This puzzle comes originally from Sam Loyd.

    On what day of the week is the following statement true?

    When the day after tomorrow is yesterday, then "today" will be as far from Sunday as that day which was "today" when the day before yesterday was tomorrow.

3 SOLVING SINGLE-VARIABLE QUADRATIC EQUATIONS

If I were you, this would be about the time that I'd be asking myself, "Okay, so what? I've learned how to add, subtract, multiply, and divide polynomials and now I'm learning how to factor them? Why?"

Well, as I told you at the beginning of the last chapter, the purposes of most of the work you're doing with polynomials are (1) to enable you to solve single-variable equations, and (2) to help you understand two-variable equations. Well, now you're ready to start doing some of that work. In this lesson we'll look at single-variable equations and in the next we'll look at two-variable equations.

Let's start with this equation:

$$x^2 + 2x = 15$$

First off, a piece of vocabulary: the equation ($x^2 + 2x = 15$) falls into the category of *quadratic equations*. So do these equations:

$$x^2 - 7x = 20$$
$$x^2 + 5x - 3 = 0$$
$$5x^2 + 3x = 25$$
$$8x^2 + 7x = -3x + 6$$

... these, on the other hand, are *cubic* equations:

$$x^3 - 4x^2 + 6x = 20$$
$$7x^3 + x = 0$$
$$-2x^3 - 16 = x^2 + 9$$

... these are *quartic* equations:

$$x^4 + 16x^3 - 3x + 8 = 0$$
$$-5x^4 - 7x^3 = 20$$
$$9x^4 + 3x^3 = 5x^2 + x$$

... and these are *quintic* equations:

$$x^5 + 6x^4 - 3x^3 + x^2 - x = 100$$
$$7x^5 + 13x^3 = 5x^2 - 28$$
$$-10x^5 + 9x^4 = -2x^3 + x^2 - 10x + 17$$

... and I could keep on going the same way (although the group that follows is usually just called "sixth-degree equations," the next one "seventh-degree equations," and so on).

1. **Based on those examples, what defines quadratic equations as a group? What defines cubic equations? What defines quartic equations? What defines quintic equations?**

For the time being, we're only going to deal with quadratic equations. (You might be interested to know that *quadratic* comes from the Latin *quadratus*, "made square," which makes a lot of sense, given what you figured out in Problem 1.) Now let's return to this equation:

$$x^2 + 2x = 15$$

Up until this point in your algebra career, you haven't had a method — other than guess and check — for solving an equation like this one. But now you're ready to figure out such a method.

2. **I'm going to ask you to take a shot at solving ($x^2 + 2x = 15$) before I give you any instructions about it. It's okay if you don't succeed, so don't spend a long time struggling with it if you're having difficulty. I'll just give you a couple of hints: You'll need to use the skills you learned in the last lesson, and ($x^2 + 2x = 15$) has two solutions.**

Okay. Maybe you figured that out on your own and maybe you didn't. Now I'll explain the process. Just in case you did manage to figure it out, I'll use a different but very similar example:

$$x^2 - 7x = 18$$

3. **I'm just going to tell you what the solutions to ($x^2 - 7x = 18$) are. They are 9 and (- 2). Check to see that those solutions make the equation ($x^2 - 7x = 18$) true.**

The first thing to do is to convert that equation to what's called *standard form*. The template for standard form is this:

$$ax^2 + bx + c = 0$$

In the case of ($x^2 - 7x = 18$), **a** is **1**, **b** is (- 7), and **c**, as you'll see in a moment, is (- 18).

4. **Convert ($x^2 - 7x = 18$) to standard form by subtracting 18 from both sides.**

5. **The next step is the one that uses your new factoring skills. Simply write a new version of ($x^2 - 7x - 18 = 0$) where you factor the left-hand side into two binomials the way that you learned to do in the last chapter.**

6. **Compare the equation you just wrote in Problem 5 with the solutions to ($x^2 - 7x = 18$) that you checked in Problem 3. This can't be a coincidence, can it? Explain how the equation you wrote in Problem 5 shows that 9 and (- 2) must be the solutions to ($x^2 - 7x = 18$).**

What you just discovered is something mathematicians call the *Zero-Product Property*. The Zero-Product Property states that if you multiply a string of numbers together and get zero as the product, at least one of the numbers that you multiplied must have been zero. Or, if you prefer to have it stated mathematically, it would look like this:

If ab = 0, then either a = 0 or b = 0

In the case of this equation:

$(x - 9)(x + 2) = 0$

... either $(x - 9)$ is zero or $(x + 2)$ is zero, so **x** must be either **9** or **(- 2)**.

The combination of your factoring skills and your understanding of the Zero-Product Property means that you are now capable of solving quite a few quadratic equations.

My complete work for solving $(x^2 - 7x = 18)$, written the way that I'd expect it to appear in your notebook, looks like this:

$$x^2 - 7x = 18$$
$$x^2 - 7x - 18 = 0$$
$$(x - 9)(x + 2) = 0$$
$$x = 9, -2$$

Notice that I've indicated the two solutions in the last step with a comma.

Solve the following quadratic equations. The first thing to do is always to change them into standard form if they aren't in it already, and this may sometimes take a few steps. Check the solutions you get by plugging them into the original equation. (It's fine to do the checking in your head.)

7. $x^2 + x - 6 = 0$

8. $x^2 + 4x - 5 = 0$

9. $x^2 + x - 12 = 0$

10. $x^2 - 8x + 15 = 0$

11. $x^2 + 9x + 18 = 0$

12. $x^2 - 2x = 15$

13. $x^2 - 11x = -30$

14. $x^2 + 14x = -40$

15. $x^2 + 2x - 20 = 3x + 10$

16. $x^2 + x + 21 = -2x^2 + 11x$

17. $-3x^2 + 6x - 17 = -4x^2 + 10x - 5$

18. $x^2 = 2(4 - x)$

19. $-12(x + 2) = x^2 + 8$

Let's talk for a minute about the number of solutions that a quadratic equation can have. It shouldn't surprise you that a quadratic equation can have two solutions, because in the past you've solved a number of simple quadratic equations, such as this one: $x^2 = 16$.

20. **If $(x^2 = 16)$ is a quadratic equation, you should be able to change it into standard form. Go ahead and do so and say what the values of a, b, and c are.**

21. **Just to see that it's possible, solve $(x^2 = 16)$ by factoring the left-hand side of the equation you wrote in Problem 20.**

Of course, you would never really solve $(x^2 = 16)$ that way. You can see right away that its solutions are **4** and **(- 4)**.

So quadratic equations can definitely have two solutions. I'll just tell you that they can't have more than two solutions. They *can* have fewer than two solutions — or at least fewer than two real solutions.

22. **Write a quadratic equation that has no real solutions and explain why it has none.**

It's also possible for a quadratic equation to have only one real solution. The most obvious example is ($x^2 = 0$), but there are others.

23. **See if you can find another quadratic equation that has only one real solution. This can be a little tricky, so don't worry if you can't, but here's a hint: Try starting with the factored form of the equation. For instance, I can see right away that this equation has two real solutions:**

$(x + 5)(x - 2) = 0$

Solve the following equations.

24. $x^2 - 6x = -9$

25. $x^2 + 4x = -4$

26. $x^2 - 5x + 20 = 5x - 5$

27. $-2x^2 + 12(x + 3) = -3x^2$

28. **Problems 24 – 27 each had only one real solution. What do the versions of those equations in standard form have in common?**

There's one more type of quadratic equation that you can solve using the skills you have so far. Here's an example in standard form:

$3x^2 - 3x - 18 = 0$

So far you haven't dealt with factoring polynomials in which the coefficient of the first term is anything other than one. But, wait, yes you have — back in Lesson 1 of this chapter. As long as all the terms in a polynomial have a common factor, there *is* something you can do.

29. **Use the skills you learned in Lesson 1 to rewrite ($3x^2 - 3x - 18 = 0$), factoring the greatest common factor out of the left-hand side. When you're done, the left-hand side of the equation should look like a number times a polynomial.**

30. **Now you can factor that polynomial the way you've been doing so far in this lesson. Go ahead and do so. Your new equation in combination with the Zero-Product Property should give you the solutions to ($3x^2 - 3x - 18 = 0$). Make sure those solutions actually work in the original equation.**

Here are a few more equations to solve. Be sure to check that your answers make the original equations true.

31. $5x^2 - 10x - 40 = 0$

32. $-4x^2 + 32x - 64 = 0$

33. $x^2 + 4x = 0$ (In this case, the greatest common factor is a variable.)

34. $2x^2 - 4x - 36 = 2x$

35. $-3x^2 + x + 70 = x - 5$

Of course, one of the points of being able to solve single-variable equations — perhaps the main point — is to be able to apply them to problem-solving situations. Problems 36 through 39 can be pretty tough, but you can handle them. You might want to work with your math group.

36. Chuckles the Amazing Rocket Dog is launched out of the dog-launching cannon. (Don't worry, folks, she'll be perfectly safe. She's been specially trained for this. Also, she lands on a giant cushion covered with dog biscuits.) The height of her flight is given by the formula ($h = 32t - 4t^2$), where h is her height in feet and t is the time that's passed in seconds. After how many seconds will she reach a height of 60 feet on the way up? After how many seconds will she again be at 60 feet on the way back down?

37. The circus is framing a memorial portrait of Chuckles. (She died, fat and peaceful, of old age.) The picture itself is 5 feet by 3 feet and they've mounted it in a very large, very garish frame so that the whole area of the portrait, including the frame, is 63 square feet. Assuming the frame is the same width all the way around, how wide is the frame? (This problem is easier to solve if you let x be the total amount that gets added to the picture — in other words, if x is twice the width of the frame. If you solve it that way, just be sure to keep in mind that x was twice the width of the frame when you check your answer. Also, you'll get two solutions but only one of them will make sense.)

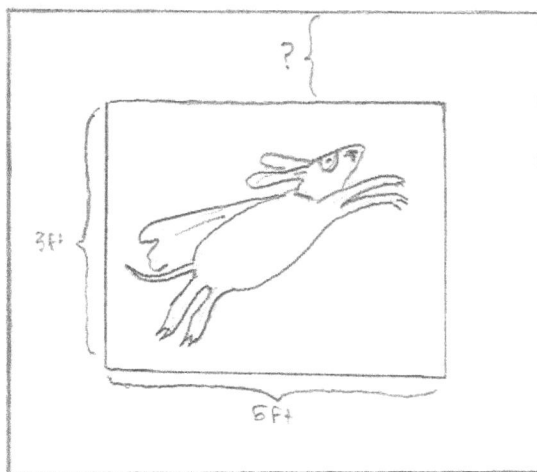

?

3ft

5ft

(not drawn to scale)

38. Lars has constructed a naked mole rat farm. (It's like an ant farm, only for naked mole rats.) He started with a large sheet of metal, twice as long as it was wide. He cut out 4' x 4' squares from the corners of the piece of metal, like so:

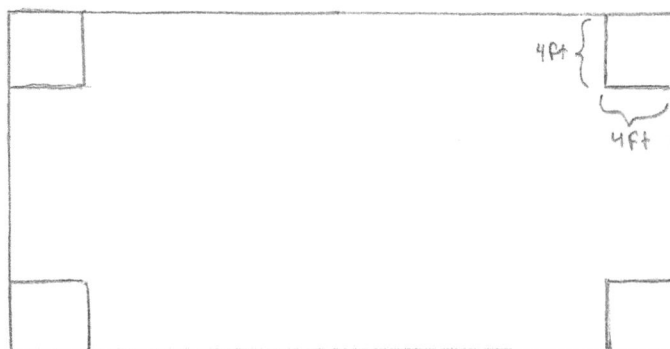

4ft

4ft

(not drawn to scale)

... and then folded up the sides so that he had a box with one open side. Then he covered that open side with a big piece of plexiglass. If the volume of the naked mole rat farm ended up being 616 cubic feet, what were the dimensions of the original sheet of metal? (Start by calling the width of the piece of metal x, then write an equation for the volume of the box based on that. You'll need all your skills from this lesson to solve the equation and again only one of the solutions will make sense. I think this is a pretty tough problem so I'd suggest working on it with your math partners.)

39. Lars also made a slide for his naked mole rats, again by folding a piece of metal. He started with a piece 24 inches wide, and the cross sectional area of the slide is 72 square inches. How deep is the slide?

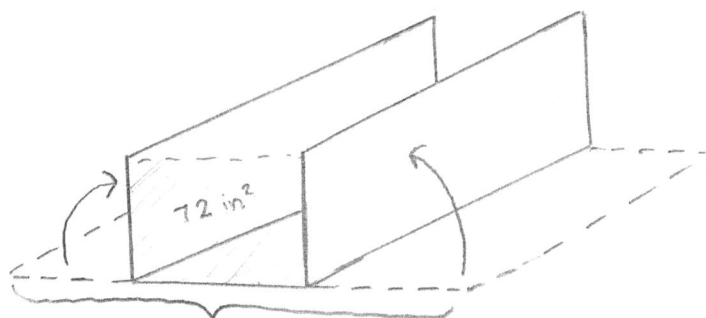

72 in²

24 in

(not drawn to scale)

40. Write your own story problem based on Problem 37, 38, or 39. You can use a similar scenario, but use different numbers. I found those problems challenging to write, and I think it's worth your while to try one. Give your problem to a classmate to solve.

41. Write a *Note to Self*. This time my suggestion for a title is a little less clunky: *Solving Quadratic Equations, Part 1*. (The "Part 1" should suggest to you that there are going to be some complications down the road.) Your Note should include at least one clearly explained example and an explanation of how the Zero-Product Property is involved in solving quadratic equations. You don't need to include a story problem.

REVIEW

1. Do the following calculation and express your answer without any negative exponents:

$(4a^3b^{-2}c)(2^{-2}a^3b^{-2}c^{-5})(3^{-1}a^3bc^2)$

2. Do the following calculation and express your answer without any positive exponents:

$\left(\dfrac{a^3b^2c}{d^5e}\right)\left(\dfrac{a^{-1}b^3c^4}{de^2}\right)$

3. Graph the equation $y = 2^{x+1}$.

4. Multiply $(x^6 - 3x^2 + 2)$ by $(2x^5 + x^3 - x)$.

5. Solve the following equation:

$\sqrt{\dfrac{x+3}{3}} = 3$

Solve the following pairs of simultaneous equations:

6. $3x + 3y = -18$
   $5x + 2y = -45$

7. $8x + 2y = 1$
   $12x - 4y = 5$

8. Remember that the speed of a car involved in a crash can be estimated using the formula $S = \sqrt{20D}$, where S is the speed in miles per hour and D is the distance in feet. Use a calculator to make a table for the values of D when S is 0, 10, 20... up to 100. (You can round the values of D to whole numbers.)

   Now make a graph of the table, remembering that D is the independent variable and S is the dependent variable. For D, one graph paper square can stand for 10 feet. For S, two squares can stand for 10 miles per hour. Connect your points with a smooth curve.

   Your graph should look like half of a kind of graph that you're very familiar with, only turned on its side. Thinking about the formula $S = \sqrt{20D}$, explain why it makes sense that your graph should have that shape. (It might help to solve the formula for D and think about what the shape of the graph would be if S were the independent variable.)

9. Solve the following equation:

$\dfrac{35}{x} + \dfrac{45}{3x} = 10$

10. This is a puzzle game invented by Lewis Carroll, author of *Alice in Wonderland*, who called it "Doublets."

The idea is to change one word into another. The rules are that you can only change one letter at a time and that each change must result in an actual new word. Let's say you were to change *more* into *less*:

more —> lore —> lose —> loss —> less

Here are three to try:

Change *milk* into *pail*.
Change *oil* into *gas*.
Change *four* into *five*.

Feel free to write some of your own for your teacher to try when he or she is checking your work!

4 GRAPHING TWO-VARIABLE QUADRATIC EQUATIONS

This will be a relatively short lesson, because in a sense it's mostly review — although it does provide a new approach to something you're already familiar with, and this approach will be useful when it comes to graphing cubic equations, quartic equations, and so on in Chapter 5.

Let's consider a two-variable quadratic equation (as you might expect, this is an equation in which the highest power to which the independent variable is raised is the second), such as:

$$y = x^2 - 2x - 8$$

First of all, you know that, like all two-variable equations, this equation has an infinite number of paired solutions that can be represented graphically on the Cartesian coordinate plane. But you know considerably more than that...

1. **At the end of *Crocodiles & Coconuts*, you learned to tell the basic shape of a graph just by looking at the equation. What is the basic shape of ($y = x^2 - 2x - 8$)? (Check your final Note to Self from *Crocodiles & Coconuts* if you need to.)**

In fact, the graphs of all two-variable quadratic equations have that same basic shape. After all, you learned to tell that the graph of an equation is a parabola if the independent variable is raised to the second power (and no higher); well, that's precisely the definition I just gave you of two-variable quadratic equations. You've graphed a lot of parabolas, but you're not used to seeing their equations in this form:

$$y = x^2 - 2x - 8$$

When you graphed parabolas in *Crocodiles & Coconuts*, their equations were in vertex form, which looks like this:

$$y = (x - 1)^2 - 9$$

2. **Vertex form is extremely useful in graphing parabolas because it tells you where the vertex is. Where is the vertex of ($y = (x - 1)^2 - 9$)? (Look back at your Note to Self on graphing parabolas from *Crocodiles & Coconuts* if you can't remember how to tell where the vertex is.)**

In fact, ($y = x^2 - 2x - 8$) and ($y = (x - 1)^2 - 9$) are two versions of the same equation.

3. **To prove that what I just said is true, take the equation ($y = (x - 1)^2 - 9$) and multiply out the $(x - 1)^2$, then simplify.**

It's pretty easy to change an equation from vertex form to a form like ($y = x^2 - 2x - 8$). In Chapter 5 you'll figure out how to go from the form ($y = x^2 - 2x - 8$) to vertex form as well. (Actually, you could probably figure it out right now if you wanted to, but I won't ask you to do it until Chapter 5.) So what does the form ($y = x^2 - 2x - 8$) tell us that vertex form doesn't?

4. **To start answering that question, factor the right-hand side of the equation ($y = x^2 - 2x - 8$) according to the method that you learned in Lesson 2 of this chapter.**

5. **Find the solutions to this single-variable equation: $0 = (x - 4)(x + 2)$**

The equation in Problem 5 is, of course, the equation that you came up with in Problem 4, with **y** replaced by zero. In other words, the two values for **x** that you found in Problem 5 are the values where **y** is equal to zero in the equation ($y = x^2 - 2x - 8$).

6. **If you think about the graph of any equation, what is the significance of the points where y is equal to zero?**

7. **Based on what you now know, you have enough information to make a pretty accurate graph of ($y = x^2 - 2x - 8$). Your answer to Problem 2 tells you where its vertex is, and your answer to Problem 5 tells you where two other significant and easy-to-find points are. Go ahead and graph it by connecting those three points to form a parabola.**

So that is what your factoring work can tell you about the graph of a quadratic equation: the points where it crosses the **x-axis**. As I say, this will come in quite handy when you start graphing higher-degree equations. For the time being I'm just going to ask you to consider a few more quadratic equations and their graphs.

8. **Use your factoring skills to find the x-intercepts of the equation ($y = x^2 + 4x$). I'll just tell you that its vertex is at (-2, -4). On the basis of those three points, graph it.**

Take a look at the graph that you just made. Notice that, between the values of ($x = -4$) and ($x = 0$), the parabola dips below the **x-axis**. In other words, the **y-values** that correspond to the **x-values** between **-4** and **0** are negative; the **y-values** for ($x < -4$) and ($x > 0$) are positive.

9. **Still thinking about the equation ($y = x^2 + 4x$), when x is positive, is x^2 positive or negative? Is 4x positive or negative?**

10. **When x is negative, is x^2 positive or negative? Is 4x positive or negative?**

One interesting perspective to take on an equation like ($y = x^2 + 4x$) is that when **x** is positive, the two terms x^2 and **4x** are reinforcing each other — they're both, in a sense, pushing the graph upward — and when **x** is negative, x^2 and **4x** are working against each other — in a sense, x^2 is pushing up on the graph while **4x** pulls down.

(You may have noticed that I've stopped leaving that extra space after the negative sign. I don't want to muck up my paragraphs with a lot of extra parentheses to make it clear to you that the sign belongs to the following term when I could just write **-4** or **-2x**, and I know I can trust you to keep in mind that there's no difference at all between a minus sign and a negative sign. From now on you'll see negative terms with or without the space, for my own convenience and to help you stay mentally nimble.)

11. Copy and fill in the following table:

x	x^2	4x
0		
-1		
-2		
-3		
-4		
-5		

12. Based on that chart, at what point does the upward push of x^2 overpower the downward pull of 4x? What does that have to do with the graph of ($y = x^2 + 4x$)?

This idea of the terms of a two-variable equation pushing up or pulling down on the graph is not something you need to master. In fact, there are probably mathematicians and math teachers who would disapprove of my using the language of "pushing" and "pulling" on the graph at all. But as I say, I think it provides an interesting way of thinking about things.

For each of the following equations, state where the x-intercepts of its graph would be:

13. $y = x^2 + 4x - 12$

14. $y = x^2 - 7x + 10$

15. $y = x^2 - 14x + 40$

16. $y = x^2 - 64$

17. $y = x^2 + 6x$

18. $y = 2x^2 - 4x - 6$ (For this one, use the technique that you figured out in the last lesson of first factoring out the greatest common factor.)

19. $y = 3x^2 - 12$

Now I'd like you to consider this equation:

$y = -x^2 - 4x - 3$

20. Based on what you know about graphing parabolas, what do you think the main difference between the graph of ($y = -x^2 - 4x - 3$) and the graph of ($y = x^2 - 4x - 3$) would be? (One way to think about it is to ask which directions the x^2 and the $-x^2$ are pushing or pulling the graph.)

As you were probably able to figure out, the graph of ($y = -x^2 - 4x - 3$) is a downward-opening parabola. It's still possible to figure out its **x-intercepts** using the techniques that you know.

21. Figure out the x-intercepts of ($y = -x^2 - 4x - 3$) by first factoring -1 out of the right-hand side and then factoring the resulting binomial. (I think this technique is a lot easier than trying to factor ($-x^2 - 4x - 3$) directly.)

For each of the following equations, state where the x-intercepts of its graph would be:

22. $y = -x^2 - 3x + 10$

23. $y = -x^2 + 11x - 30$

24. $y = -4x^2 + 16x + 48$

25. $y = -5x^2 + 20x + 105$

There's just one last thing I'd like to look at in this lesson.

26. Graph the equation $y = (x - 2)^2 + 2$.
   (Remember that it's easy to find the y-intercept of an equation by plugging in zero for x. You can use the vertex, the y-intercept, and the point that mirrors the y-intercept on the other half of the parabola to make the graph.)

27. Now convert the equation ($y = (x - 2)^2 + 2$) to a form like ($y = x^2 - 2x - 8$) by multiplying out the $(x - 2)^2$ and simplifying.

28. Look at the graph you made in Problem 26. What does that graph tell you about the polynomial on the right-hand side of the equation you got in Problem 27?

29. If you can't factor the right-hand side of an equation like, for example, ($y = x^2 + 4x - 11$), what does that tell you about its graph?

30. I was a little torn about whether or not to ask you to write a **Note to Self** on the material in this lesson, because I know there will be more on this subject in Chapter 5. But since there's going to be a Chapter 3 test and I know that a lot of students like to use their Notes to study for tests, I think it would be a good idea to write one. It should remind you how to go about **finding the x-intercepts of quadratic equations**.

REVIEW

Graph the following equations:

1. $\dfrac{y}{-4} + \dfrac{x}{5} = 1$

2. $\dfrac{y^2}{12} + \dfrac{x^2}{15} = 1$

   (Estimate the values of the x- and y-intercepts. You can do this since you know how to estimate square roots.)

3. Divide $(2x^3 - 13x + 10)$ by $(x - 2)$.

4. Simplify the expression $\sqrt{75} + \sqrt{48} + \sqrt{8}$.

5. Millicent enjoys lounging on a beanbag filled with lima beans, black beans, pinto beans, and white beans in a ratio of $2 : 8 : 3 : 5$. To the nearest tenth of a percent, what percent of the beans are pinto beans?

6. For the following fraction, what values of the variables are prohibited?

   $\dfrac{3}{xy + z}$

7. Find the greatest common factor of $126x^2y^2$, $60xy^3$, and $210x^3y$.

8. Simplify the expression $3 - |10 - 20| - (2^2 - 5)^2$.

9. Romeo is two-thirds of Mercutio's age. In ten years, if either of them were going to live that long, Romeo would be fourth-fifths of Mercutio's age. How old are they now?

10. Here's a riddle from BrainBashers.com:

    The person who makes it doesn't need it.
    The person who buys it doesn't need it for himself.
    The person who uses it doesn't know it.
    What is it?

5 FURTHER FACTORING OF SECOND-DEGREE POLYNOMIALS

It's now time to delve further into factoring polynomials. So far, as you know, you've mainly factored polynomials whose first terms have coefficients of **1**. At this point, you've gotten quite good at that sort of factoring. The only situations in which you've factored polynomials with first-term coefficients other than **1** have been when all the terms have a greatest common factor.

You're still going to be doing exactly the same thing as you did in Lesson 2: factoring polynomials (mostly trinomials) into two binomials. But when the first coefficient of the polynomial you're factoring is something other than **1** and the terms have no greatest common factor, you need a new technique.

Let's take a look at why that is.

1. **Multiply (3x + 1) by (2x + 5).**

So you ought to be able to factor ($6x^2 + 17x + 5$). After all, we know that ($3x + 1$) and ($2x + 5$) are factors of ($6x^2 + 17x + 5$). But if you try to use the technique that you learned in the Lesson 2, you'll pretty quickly see that it doesn't work: there are no two numbers that can be multiplied to give you **5** and added to give you **17**. Also, if you look at your work for Problem 1, it's clear that the "**3**" and the "**2**" in ($3x + 1$) and ($2x + 5$) are involved in determining the "**17x**" of ($6x^2 + 17x + 5$).

It's theoretically possible to factor ($6x^2 + 17x + 5$) in your head almost the way you factored the polynomials in Lesson 2, but it's really, really hard — you just have to keep track of too much. The best way to do it is to use a box, as you did for multiplying and dividing polynomials in the last chapter.

As I did in the last chapter, I'm going to walk you through an example while you do another problem alongside me. I'm going to factor ($8x^2 + 16x - 10$) while you factor ($6x^2 - x - 15$). As was the case in the last chapter, the sample problem will take a while to get through, but you'll get quicker with practice during the subsequent problems. Also, I'm giving you a relatively challenging sample problem so that you can practice all of the steps.

First I'll make a two-by-two box (for right now the polynomials will all factor into two binomials — later I'll give you a few harder ones) and fill in the two things I'm sure about:

$$8x^2 + 16x - 10$$

I like to write the polynomial that I'm factoring up above my work, leaving some space between the polynomial and the box. I'd suggest that you do the same.

2. **Make a similar box for your problem.**

I also know that I'm definitely going to have to multiply two things in order to get **8x²** and two things in order to get **-10**. There are several choices in each case, and I like to write them down so I can test them in an efficient fashion and keep track of what I've tested. At least for now, I strongly suggest that you do something similar. I keep track of them like this:

$$8x^2 + 16x - 10$$

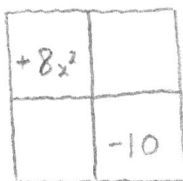

Notice that I've tried to leave some room between **(8x)(x)** and the left-hand side of my box. I'll be writing stuff in there.

Also notice that I've only listed half of the possible factor sets for **8x**. I did not include **-8x** and **-x** or **-2x** and **-4x**. The reason for this has to do with what you learned at the very end of Lesson 2. Actually, (**8x² + 16x - 10**) has two sets of factors that are opposites of one another, like (**x - 5**)(**x - 7**) and (**- x + 5**)(**- x + 7**) from the end of Lesson 2. But since we're only looking for the "positive" factors (and anyhow, the "negative" ones are easy to find once you've found the "positive" ones), I've left **-8x**, **-x**, **-2x**, and **-4x** off my list.

3. **Make similar lists for your problem.**

Now I start testing combinations of those factor pairs. I'll try **8x** and **x** along with **10** and **-1** first. There are two possible ways they could go together. The first would look like this:

$$8x^2 + 16x - 10$$

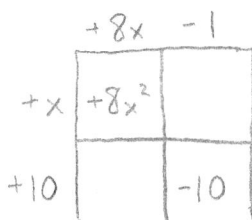

... and I'll test to see whether that combination works:

$$8x^2 + 16x - 10$$

... which it doesn't, because **80x + - x** is definitely not **16x**, which is the middle term of (**8x² + 16x - 10**). In fact, it's often the case that you can do that last step of checking in your head: I can see that (**8x**)(**10**) + (**x**)(**-1**) is not going to give me **16x** without actually writing the **80x** and the **-x** into the compartments, but I've done it here in order to show you my thinking.

4. **Choose a set of factors for your problem and test them in a similar way. It's possible that you'll guess right the first time and if you do, that's great. But I'm kind of hoping you won't so you'll have to take a couple more steps. Even if you do guess right, follow along with me for the rest of my process.**

I'll erase the **80x** and the **-x** from my work and try again. I want to be systematic, so I'm going to try the other possible combination of **8x, x, 10,** and **-1**. It looks like this:

$$8x^2 + 16x - 10$$

Well, **10x** plus **-8x** is **2x**. But I'm looking for **16x**. Closer than the last time but still, as the saying goes, no cigar. (This saying supposedly comes from the days when fairground booths gave out cigars as prizes rather than giant stuffed animals and the like.)

I'll keep testing.

Once I've tried **8x** and **x** combined in all of the possible ways with **-10** and **1**, with **5** and **-2**, and with **-5** and **2**, and found that none of those possible combinations work (they don't; trust me on this), I can cross **8x** and **x** off my list on the left and start testing **2x** and **4x**.

Eventually I'll find a combination that works and I'll record my final factoring. Ultimately, my page will look something like this:

$$8x^2 + 16x - 10 = (2x + 5)(4x - 2)$$

5. **Follow a similar process until you find the factors of (6x² - x - 15), assuming you didn't get them right on the first guess. The important thing is to be systematic about it: keep track of which combinations you've tried so you don't wind up trying things over and over. Be sure to check that your final answer is correct by doing the multiplication.**

As you can see, factoring polynomials like the one you just worked on is essentially a process of guess and check — with a system for keeping track of your guesses. Sometimes the process can be rather long and involved, especially if the numbers that you're working with happen to have lots of factors. The fact of the matter is that there are no great short-cuts available — sometimes you just have to test all of the potential factors. However, you can streamline the process a little bit by being smart about which sets of factors are most likely and testing those first.

In the problem I was working on, I needed my two middle terms to add up to **16x**. There was basically zero chance that the combinations that involved multiplying **8x** by either **10** or **-10** would work: the monomial that I would need to add to **80x** or **-80x** in order to get to **16x** would just be too big. I should have chosen to start by testing other combinations.

In the problem you were working on, you were aiming to have your two middle terms add up to **-x**. That meant that the two terms you were adding together had to be very nearly opposites of one another. So you probably should have started with combinations that were not very extreme. For instance, the combinations that involve multiplying **6x** by **15** and **-15** would probably have been a silly place to start.

If you use this kind of logic, it may help you to factor some of the polynomials in this next problem set more efficiently. I think you'll find that some of them are easier than your example problem; I gave you a relatively tough one to start with. These are hard enough, though, that I will not give you any prime polynomials — it would be pretty frustrating to try every possible combination and have none of them work — but do realize that it's perfectly possible that at some point you'll encounter a prime polynomial of this kind.

Factor the following polynomials:

6. $3x^2 + 11x + 6$ (As you set up this problem, think about the signs of the three terms — you'll find you only have to test two factor pairs for the third term instead of four.)

7. $5x^2 - 12x + 4$ (Again, you only need to test two factor pairs for the third term.)

8. $2x^2 + 21x - 11$ 9. $10x^2 - 7x + 1$

10. $9x^2 - 12x - 5$ 11. $2x^2 + 7x + 6$

12. $4x^2 - 25x - 21$ 13. $4x^2 - 27x + 18$

14. $4x^2 - 17x + 18$

15. $16x^2 - 1$ (This is the easiest problem so far. You don't even really need to set up a box. You just need to think about your work in Lesson 2 and remember one specific factoring situation...)

16. $10x^2 - 31x + 15$ 17. $5x^2 - 2x - 3$

18. $2x^2 - x - 3$ 19. $25x^2 - 16$

20. $-2x^2 + x + 6$ 21. $-4x^2 - 4x + 3$

22. $8x^2 + 4x - 24$

23. When I wrote Problem 22, I thought it would be pretty difficult because -24 and $8x^2$ have so many possible factor pairs. However, I discovered while I was editing my work (for the third time!) that there are three possible correct answers, so it probably wasn't as hard as I'd thought it would be after all. You might have gotten $(8x - 12)$ $(x + 2)$ or $(2x - 3)(4x + 8)$ or $(4x - 6)(2x + 4)$. Any of those combinations would give you $(8x^2 + 4x - 24)$. Your job is to examine those three combinations and show that they can all be simplified further in order to get exactly the same thing in each case. (Hint: You'll need to use the skill you learned in Lesson 1 of this chapter.) Based on what you just did, what was the very first step you and I should have taken in factoring $(8x^2 + 4x - 24)$? (For bonus points you could go back through this lesson and find the other place where I made this same mistake.)

I hope you realized that Problems 15 and 19 both fell into the category of the difference of squares that you studied in the last chapter. They should have been easy to factor.

There's one other special situation besides the difference of squares that I'd like you to recognize because it will sometimes make your factoring work easier.

You did problems like these in Lesson 3 of the last chapter, but for practice, square the following binomials to get trinomials:

24. $(2x + 6)^2$

25. $(x - 3)^2$

26. $(3x + 2)^2$

The trinomials that you got can be called *perfect trinomial squares* because they are the squares of binomials. All I really want you to notice is that the first and the last terms of those trinomials are perfect squares. If you're asked to factor a trinomial the first and last terms of which are perfect squares, the very first thing you should do is test to see if it's the square of a binomial. There are cases where it won't be, but it's still a very good guess and will often cut short the factoring process.

Factor the following trinomials:

27. $4x^2 - 12x + 9$

28. $16x^2 + 40x + 25$

29. Find the values of the trinomial $(x^2 + 2x + 1)$ when x is equal to 0, 1, 2, 3, and 4. What do you notice about those values? Factor $(x^2 + 2x + 1)$ and use that factoring to explain what you've noticed about those values. What would you predict about the values of the trinomials $(4x^2 - 12x + 9)$ and $(16x^2 + 40x + 25)$ (from Problems 27 and 28) when you plug x-values into them? Plug a couple of x-values into each trinomial to see whether your prediction was correct.

There's one more situation involving the difference of squares that I'd like to show you. Consider, for example, this expression:

$(x + 2)^2 - 9$

Notice that it's the difference of two squares: $(x + 2)^2$ and **9**. That means that it can be factored like so:

$[(x + 2) - 3][(x + 2) + 3]$

30. Verify that you can multiply $[(x + 2) - 3][(x + 2) + 3]$ and get $[(x + 2)^2 - 9]$. When you multiply the expression $(x + 2)$ by itself, just write the result as $(x + 2)^2$. Don't multiply it out.

The expression $[(x + 2) - 3][(x + 2) + 3]$ can then be simplified to $(x - 1)(x + 5)$. I just combined the constants in those expressions: in other words, I added **2** and **-3** to get **-1** and **2** and **3** to get **5**.

Use a similar technique to factor the following expressions as the difference of squares. Simplify the results as much as possible.

31. $(x - 5)^2 - 16$

32. $(3x + 2)^2 - 1$

33. $25 - (x + 3)^2$

34. $(2x - 5)^2 - (3x + 1)^2$

35. $(x + y)^2 - 144$

Just as in Lesson 2, the techniques that you've learned in this lesson can help you factor some even more complex polynomials, including ones that have more than one variable and some that are higher than second degree. All of the polynomials in the next set can be factored using the box method that you've been practicing.

36. $3x^2 + 7xy + 2y^2$

37. $2xy + 6x + 5y + 15$

38. $3x^4 + 11x^2 + 10$

39. $8x^{10} + 2x^5 - 3$

40. This method of factoring will only work on very specific sorts of polynomials of higher than second degree. Look at the exponents for Problems 38 and 39. What pattern do they fall into?

41. It's time for a *Note to Self*. One of a series, really. This one could be *Factoring Second-Degree Polynomials: Polynomials with First-Term Coefficients Other than One*. That may be my worst suggestion for a Note to Self title yet! At any rate, this Note should include instructions for how to do the factoring work that you learned in this lesson. Pick and explain one good example.

REVIEW

1. Graph the equation $y = |x + 1|$.

2. Graph the simultaneous inequalities $y \leq -x^2 + 3$ and $y \geq x^2 - 3$.

3. Estimate the value of the expression $\sqrt{32} - 2\sqrt{7}$.

4. Convert $\sqrt{\dfrac{7x^3}{3y}}$ to simplest radical form.

5. Add $(3x^5 - 7x^3 + x^2 - 9)$ to $(-6x^5 + 4x^4 + 10x^2 - 3)$.

6. Linus is participating in a self-control experiment. He's left alone in a room with 40 marshmallows. After 8 minutes, the marshmallows are all gone. If there's a linear relationship between the passage of time and the number of marshmallows left (and let's go ahead and assume that Linus has stuffed his face at a steady rate and without any initial hesitation), write an equation in two-intercept form for how many marshmallows there are depending on how much time has passed.

7. Simplify the following expression and write it without any negative exponents:

$$\dfrac{\dfrac{3m^3 n^{-2} q^5}{r^6 t^{-3}}}{\dfrac{6r^{-2} t^{-6}}{m^7 n^{-2} q}}$$

8. Solve the following equation:

$$\dfrac{2m}{3} + \dfrac{3}{2} = 2(m-8) - 2\tfrac{1}{2}$$

9. Princess Ripley needs to leave her palace and ride to the rescue of the prince in the enchanted tower. It's imperative that she arrive exactly on time in order to break the enchantment. (And she's fairly certain the prince is worth this effort.) She's calculated that if she rides her trusty charger, Engelbert, at 10 miles an hour she'll get there an hour late and if she rides at 15 miles an hour she'll get there an hour early. How far is it from her palace to the enchanted tower?

10. Paul the Octopus was predicting the outcome of a soccer tournament. (Strangely enough, although many things in these textbooks are made up, Paul is not. He was a real octopus in Germany who correctly predicted the outcome of all seven of his country's games in the 2010 FIFA World Cup as well as the final match between Spain and the Netherlands. The rest of the problem is made up, though.) Here are Paul's predictions:

1. Armenia will finish ahead of Denmark but behind Bulgaria.
2. Bulgaria will finish ahead of Cameroon but behind Hungary.
3. Hungary will finish ahead of Finland.
4. Finland will finish ahead of Ghana.
5. Ghana will finish ahead of Indonesia.
6. Indonesia will finish ahead of Jamaica.
7. Hungary will finish behind Kyrgyzstan.
8. Finland will finish behind Denmark.
9. Denmark will finish ahead of Ghana.
10. Ghana will finish behind Estonia.
11. Estonia will finish ahead of Denmark.
12. Cameroon will finish ahead of Estonia.
13. Estonia will finish behind Bulgaria.
14. Cameroon will finish ahead of Ghana.
15. Armenia will finish behind Estonia.

What final rankings did Paul predict?

6 APPLICATIONS IN EQUATION-SOLVING & GRAPHING

In my opinion, the last lesson was pretty tough. To make up for it, this lesson is going to be very brief. You'll have to do some of the same kind of factoring that you did in the last lesson, but not a ton of it, and I promise I won't give you any insanely hard ones with lots of factor pairs. I just want to bring the skills that you learned in Lesson 5 to bear on the stuff you were working on in Lessons 3 and 4. I probably could have just stuck this lesson's material at the end of Lesson 5, but honestly I just wanted you to be able to pause and take a breath.

As you know, the factoring skills that you've been learning in this chapter allow you to solve certain kinds of single-variable equations and help you graph certain kinds of two-variable equations.

1. In order to refresh your memory, find the solutions to $(x^2 + 2x - 35 = 0)$.

2. And name the x-intercepts of the graph of $(y = x^2 - 13x + 30)$.

The work that you did in the last lesson expands your repertoire of solvable and graphable equations.

3. Use the factoring methods you learned in the last lesson to find the solutions to $(2x^2 + 5x - 3 = 0)$.

4. Notice that one of the solutions to $(2x^2 + 5x - 3 = 0)$ is a fraction. If you haven't already done so, just this once I want you to check that both of the solutions you found make $(2x^2 + 5x - 3 = 0)$ true. In this case, the fraction isn't too hard to plug in — you just have to think about what happens when you multiply a fraction by itself — but some of the other ones you're going to do would be hard to check, so from here on out when you get a fractional answer, you can just check it with your math partners instead of plugging it into the equation.

5. Look back at the work you did in Problem 3. What are the x-intercepts of $y = 2x^2 + 5x - 3$?

Find the solutions of the following equations:

6. $15x^2 + 19x + 6 = 0$

7. $9x^2 - 15x - 14 = 0$

8. $63x^2 - 5x - 2 = 0$

9. $16x^2 + 14x = 15$

Name the x-intercepts of the graphs of following equations:

10. $y = 3x^2 - 13x - 10$

11. $y = 8x^2 + 6x - 9$

12. $y = 100x^2 - 1$

13. $y = 12x^2 - 37x - 10$

14. Lars is now making a table for his naked mole rat house. (Don't ask me why naked mole rats would need a table.) The first square tabletop he cut was too small. He made each edge of his new tabletop half a foot longer than the edges of the old tabletop. The surface area of the new tabletop is four times that of the old top. What were the dimensions of the original tabletop?

(I'd recommend doing this challenging problem with partners. Try letting the length of one side of the old tabletop be x. After you write your equation and square the expression for the new tabletop, you'll want to multiply the entire equation by something to get rid of the fraction, then keep solving. At least, that's the way I did it. Maybe you'll find an easier way. But try to solve it algebraically, okay?)

That's it! That's the whole lesson. You don't even have to write a Note to Self. Just be aware that you may have to use your skills to solve equations and find x-intercepts on the chapter test.

REVIEW

Factor out the greatest common factor of the terms of the following polynomials:

1. $12x^3 - 9x^2 + 6x - 15$

2. $8x^5 + 10x^4 - 2x^3 + 6x^2$

3. $15x^3y^2 + 25x^2y^2 - 5xy^2$

Partial factor the following polynomials. (For example: $14x^2 - 7x + 2 = 7x(7x - 1) + 2$)

4. $9x^3 - 6x^2 + 5$

5. $4xy^3 + 8y^3 + 7$

Factor each of the following polynomials into the product of two binomials:

6. $x^2 - 11x + 10$

7. $x^2 + 6x - 40$

8. $x^2 - 3x - 54$

9. $x^2 - 121$

10. $x^4 - 16x^2 + 64$

11. $xy + 7x - 3y - 21$

Solve the following equations:

12. $x^2 - 9x + 14 = 0$

13. $x^2 + 10x = 24$

14. $3x^2 + 7x + 10 = 2x^2 - 6x + 58$

15. Chuckles's admirers have acquired a square plot of land that they intend to turn into a play area for their beloved Rocket Dog. Chuckles requires a lovely 2 meter-wide path down one side of her play area, so only 48 m² are left to be covered in grass. What are the dimensions of the whole play area? (Please try to solve this problem algebraically. There's a hint at one way to approach it in the illustration.)

48 m²

2 m

(not drawn to scale)

For each of the following equations, state what the x-intercepts of the graph would be:

16. $y = x^2 - x - 56$

17. $y = x^2 - 25$

18. $y = 3x^2 - 9x - 30$

19. $y = -x^2 + 6x + 7$

Factor the following polynomials into the products of two binomials:

20. $6x^2 - x - 15$

21. $4x^2 + 21x + 20$

22. Factor the following expression according to this method:
$(x - 5)^2 - 16 = [(x - 5) + 4][(x - 5) - 4] = (x - 1)(x - 9)$

$(x - 3)^2 - 9$

Solve the following equations:

23. $7x^2 + 27x - 4 = 0$

24. $-10x^2 + x = -18x + 6$

25. What are the x-intercepts of the graph of $y = 12x^2 - 5x - 2$?

4

POLYNOMIALS IN RATIONAL EXPRESSIONS

1 USING YOUR FACTORING SKILLS TO WORK WITH FRACTIONS

This short chapter is going to feel like a little bit of a digression. It is not focused directly on what I consider to be the main purposes of algebra: using single-variable equations as problem-solving tools; using multi-variable equations to understand the world; or proving mathematical propositions. But it is material that you may still enjoy and that ought to be part of a basic algebra curriculum. It has to do — as you might expect from the title — with fractions and polynomials.

You've worked quite a bit with algebraic fractions in the past, but your new knowledge of polynomials allows you to tackle even more complex algebraic fractions. First I'll ask you to do a few review problems.

Simplify the following fractions:

1. $\dfrac{4x^3}{x^2}$

2. $\dfrac{3m^2n^5}{12mn^6}$

3. $\dfrac{5xy}{10x^2y^2}$

4. What are you actually doing when you simplify $\dfrac{4x^3}{x^2}$ to 4x? In other words, why are you allowed to do that simplification?

5. In the fraction $\dfrac{4x^3}{x^2}$ there is a value that x is absolutely not allowed to be equal to.

What is that value?

Remembering that the denominator of a fraction can never be equal to zero, what value or values for the variables are forbidden in the following fractions?

6. $\dfrac{5x}{x+1}$

7. $\dfrac{3}{x-4}$

8. $\dfrac{4x^2}{(x+2)(x-2)}$

9. $\dfrac{8}{x-y}$

I'd like you to keep in mind that any time there are variables in the denominator of a fraction, values of those variables that would make the denominator equal to zero are forbidden. It's not something that really affects the calculations you'll be doing in this book, so you don't need to worry about it, but you should be aware that it's true. I'm going to assume that you've got the idea so I don't have to keep pointing out that certain values are prohibited every time I give you a new fraction to deal with.

I'm sure you recall that you can create equivalent fractions by multiplying both the numerator and the denominator of a fraction by the same thing. In other words, $\frac{2}{3}$ is equal to $\frac{(2)(2)}{(2)(3)}$ or $\frac{4}{6}$. It's also equal to $\frac{8}{12}$ (multiply both numerator and denominator by **4**) or $\frac{22}{33}$ (multiply by **11**) or $\frac{44}{66}$ (by **22**) or, indeed, an infinite number of other options. Using the same logic, the fraction $\frac{x}{3}$ is equal to $\frac{2x}{6}$ (multiply both numerator and denominator by **2**), $\frac{5x}{15}$ (multiply by **5**), and $\frac{10x}{30}$ (by **10**). If you multiply both numerator and denominator of $\frac{x}{3}$ by x, you'll find that it's also equal to $\frac{x^2}{3x}$, and if you multiply them both by **y**, it's equal to $\frac{xy}{3y}$. (Actually, you could get into a philosophical debate about whether $\frac{x}{3}$ is really equivalent to $\frac{x^2}{3x}$. What makes it debatable is that in the new version of the fraction, the value **x = 0** is forbidden, while in the original fraction **x = 0** is just fine. However, we'll set this debate aside and say, for the purposes of this textbook, that $\frac{x}{3}$ and $\frac{x^2}{3x}$ are equivalent fractions.)

Now consider the fraction $\frac{5x}{x + 1}$. First off, note that its denominator is a polynomial. Fractions that contain polynomials function just like any other fractions. So, in this case, you can find equivalent fractions for $\frac{5x}{x + 1}$ as long as you keep in mind that you need to multiply the *entire* denominator by whatever you choose to multiply both numerator and denominator by. In other words, you'll need to apply a certain famous rule that I've talked about and you've used tons of times and so I'm not going to name it now because you ought to know what it is.

10. **What rule am I talking about?**

11. Use that rule to find four equivalent fractions for $\frac{5x}{x+1}$.

Great. I want to point out three possible equivalent fractions for $\frac{5x}{x+1}$ that might not have occurred to you. Here's one:

$$\frac{5x}{x+1} = \frac{(x)(5x)}{(x)(x+1)} = \frac{5x^2}{x^2+x}$$

Here's another:

$$\frac{5x}{x+1} = \frac{(x-1)(5x)}{(x-1)(x+1)} = \frac{5x^2-5x}{x^2-1}$$

And here's a third:

$$\frac{5x}{x+1} = \frac{(x+3)(5x)}{(x+3)(x+1)} = \frac{5x^2+15x}{x^2+4x+3}$$

What I'd like you to notice is that it probably wouldn't be obvious at first glance that

$\frac{5x^2}{x^2+x}$ or $\frac{5x^2-5x}{x^2-1}$ or $\frac{5x^2+15x}{x^2+4x+3}$ is equivalent to $\frac{5x}{x+1}$. In fact, to see that they are all

equivalent to $\frac{5x}{x+1}$, what you would have to do is factor them. And simplifying fractions

by factoring is a good deal of what you'll be doing in this lesson.

A fraction is simplified to its lowest common terms when its numerator and denominator have no common factor other than **1**. These are some fractions that have been simplified to their lowest common terms:

$$\frac{2}{3} \quad \frac{5}{7} \quad \frac{13}{8}$$

And so are these:

$$\frac{x^2}{7} \quad \frac{x^2}{y} \quad \frac{x+1}{x-1} \quad \frac{5x-2}{x-1}$$

When I first introduced you to the idea of polynomials, I told you that they were really a new way of thinking about something familiar and that part of the point was that you should start thinking of polynomials as units instead of as strings of addition and subtraction. In order to use your new factoring skills to simplify algebraic fractions, you definitely need to think of polynomials as units.

12. Explain why $\frac{2(x+1)}{3(x+1)} = \frac{2}{3}$. (Hint: It is exactly the same reason that $\frac{2x}{3x} = \frac{2}{3}$ or $\frac{2 \cdot 5}{3 \cdot 5} = \frac{2}{3}$.)

Based on your work in Problem 12, simplify the following fractions:

13. $\dfrac{6(x-4)}{12(x-4)}$

14. $\dfrac{x+1}{2(x+1)}$

15. $\dfrac{3(x-5)}{x-5}$

I hope those weren't too tough, but the trick is that usually the polynomials won't already have been factored for you — you'll need to do the factoring and then simplify the fractions. For instance, consider the simple case of this fraction:

$\dfrac{3x-15}{x-5}$

If you factor the greatest common factor out of the numerator, you get the fraction from Problem 15, which can then be simplified. Obviously, this will get a little trickier when you're dealing with fractions like the ones I mentioned earlier:

$\dfrac{5x^2}{x^2+x}$ $\dfrac{5x^2-5x}{x^2-1}$ $\dfrac{5x^2+15x}{x^2+4x+3}$

... but you can handle it. You've done a lot of factoring, and most of the work in this lesson will be the simpler kind of factoring where the first-term coefficient is **1**.

But before I give you a long set of fractions to simplify, let me give you one word of caution: Don't try to simplify a fraction more than it can be simplified. For example, consider this fraction:

$\dfrac{3+x}{3-x}$

Though at first glance you might suppose that you ought to be able to be simplify this fraction,

you can't. Specifically, many people see those **3's** and think that $\dfrac{3+x}{3-x}$ ought to simplify

to $\dfrac{x}{-x}$ and further to **-1**. But it's pretty easy to prove that $\dfrac{3+x}{3-x}$ is not the same as $\dfrac{x}{-x}$ or **-1**.

16. **In order to prove that** $\dfrac{3+x}{3-x}$ **is not equal to** $\dfrac{x}{-x}$ **or -1, try replacing x with 1, with 6,**

and with 0, and in each case show that $\dfrac{3+x}{3-x}$ **is not the same as** $\dfrac{x}{-x}$ **or -1.**

So don't be tempted to oversimplify. The key, once again, is to think of polynomials as units. In the fraction $\frac{3+x}{3-x}$, you are really not dealing with **3's** or with **x's**, but with the polynomials **(3 + x)** and **(3 - x)**, which have no common factors. To risk beating a dead horse, let me give you one more example. You can make the following move:

$$\frac{(x+5)(x-2)}{(x+5)(x-6)} = \frac{x-2}{x-6}$$

... because **(x + 5)** is a factor of both the numerator and the denominator. Whatever **(x + 5)** is, it's the same throughout the equation and when you divide it by itself you get **1**, which is why it disappears. But you can't simplify $\frac{x-2}{x-6}$ any further because neither **(x - 2)** nor **(x - 6)** has any factors besides itself and **1**, let alone any common factors.

Okay, I'm done with my examples.

Use your factoring skills to simplify the following fractions as much as possible:

17. $\dfrac{2x+6}{3x+9}$

18. $\dfrac{2x-6}{4x+12}$

19. $\dfrac{x^2-9}{4x+12}$

20. $\dfrac{xy-10y}{x+10}$

21. $\dfrac{x^2+2x-8}{x^2-2x}$

22. $\dfrac{3x^2+12x-15}{6x^2-6x}$

23. $\dfrac{2x^2+2x-12}{3x+12}$

24. $\dfrac{10x^2-40}{x^2+4x+4}$

25. $\dfrac{3x^2-12}{5x^3-30x^2+40x}$

26. $\dfrac{x^2-x-30}{x^2-9x+18}$

27. $\dfrac{x^2-x-2}{x^2-7x+10}$

28. $\dfrac{3x^2+14x+8}{3x^2-10x-8}$

29. $\dfrac{x^2-y^2}{3x^2y+6xy^2+3y^3}$

30. Write a **Note to Self** about *simplifying fractions with polynomials.* You don't need to explain how to factor polynomials because you've done that in previous Notes, but you should explain, using examples, how to simplify fractions once you've factored the polynomials involved.

REVIEW

1. Graph the equation $x = |y|$ and explain why it cannot be a function.

2. Solve the inequality $-2x \geq 2x - 6$.

Apply the appropriate Laws of Exponents to the following expressions:

3. $\left(\dfrac{3}{y^3}\right)^{-2}$

4. $(5m^3)^2(2m^2)^3$

5. Graph the equation $16x^2 + y^2 = 16$.

Given that the made-up symbol $\tilde{\ }$ is defined like this: $x \mathbin{\tilde{\ }} y = 11(x + 2y) - 11x - 21y$, solve the following problems:

6. $3 \mathbin{\tilde{\ }} 5$

7. $7.25 \mathbin{\tilde{\ }} -4$

8. Find the least common multiple of 90 and 300.

9. Three witches are mixing up a potion. Essence of Newt Eye can be obtained at 6 shillings an ounce. Frog-Tongue Tincture costs 10 shillings an ounce. If the first step in their potion-making is to mix those two ingredients, and the resulting mixture amounts to 15 ounces and is worth 124 shillings, how much of each did they use?

10. Here are three more Doublets:

Turn *mice* into *rats*.
Turn *ink* into *pen*.
Turn *tears* into *smile*.

2 ADDING & SUBTRACTING FRACTIONS WITH POLYNOMIALS

You should learn to add, subtract, multiply, and divide fractions that include polynomials. In this lesson we'll cover addition and subtraction, and in the final lesson of the chapter we'll cover multiplication and division as well as a particular subject called *complex fractions*. Working with fractions that contain polynomials follows precisely the same principles as working with fractions that don't contain polynomials, so first I'll ask you to do some quick review problems.

Do the following problems:

1. $\dfrac{3}{7} + \dfrac{1}{7} + \dfrac{2}{7}$

2. $\dfrac{4}{11} - \dfrac{2}{11}$

3. $\dfrac{1}{6} + \dfrac{3}{8}$

4. $\dfrac{2}{9} - \dfrac{1}{3}$

In order for fractions to be added or subtracted, they need a common denominator. Remember that the word *denominator* comes from the Latin for "to name," so the denominator tells what kind of fraction you're working with; *numerator* comes from the Latin for "to count," so the numerator tells you how many of those fractional pieces you're dealing with.

So, if you're adding $\dfrac{2}{5}$ and $\dfrac{1}{5}$, you end up with $\dfrac{3}{5}$. Similarly, $\dfrac{4}{x} + \dfrac{6}{x} = \dfrac{10}{x}$.

The main challenge in adding or subtracting fractions with polynomials is, once again, that you have to think of the polynomials as units.

Try this addition problem:

5. $\dfrac{3}{x+1} + \dfrac{5}{x+1}$

I hope you were able to conclude that the denominator of your final fraction in Problem 5 should be **x + 1**, just as the denominator of your final fraction in Problem 1 remained **7**. When you do the next set of problems, the one thing to be careful of is that you should simplify the resulting fractions according to the methods that you learned in the last lesson — sometimes the final fraction can be simplified even when the original fractions cannot. The same thing is sometimes true of problems without polynomials, such as this one:

6. $\dfrac{2}{5} + \dfrac{3}{5}$

Do the following addition and subtraction problems. Simplify the resulting fractions when possible.

7. $\dfrac{x+2}{x-3} + \dfrac{x-1}{x-3}$

8. $\dfrac{5}{6x+2} - \dfrac{3}{6x+2}$

9. $\dfrac{3x}{3x-2} - \dfrac{2}{3x-2}$

10. $\dfrac{x}{2y} - \dfrac{x+2}{2y}$

11. $\dfrac{3}{x^2-9} + \dfrac{x}{x^2-9}$

12. $\dfrac{x+3}{x-2} - \dfrac{x+8}{x-2}$

(Make sure you subtract the entire numerator of the second fraction from the first.)

13. $\dfrac{x+5}{x-5} - \dfrac{x-5}{x-5}$

14. $\dfrac{x^2+1}{x^2} - \dfrac{1}{x^2}$

15. $\dfrac{x^2}{x-1} - \dfrac{x}{x-1}$

16. $\dfrac{x^2-3}{x+2} - \dfrac{1}{x+2}$

17. $\dfrac{x^2}{x-y} - \dfrac{y^2}{x-y}$

18. $\dfrac{5x^3}{5x-1} - \dfrac{x^2}{5x-1}$

19. $\dfrac{x}{x^2+x-6} - \dfrac{2}{x^2+x-6}$

Of course, oftentimes the fractions that you are adding and subtracting do not already have a common denominator — instead, you need to find one. It's simple enough to do. Here's a problem for you to warm up on, taken straight out of *Jousting Armadillos*:

20. $\dfrac{3}{5x} - \dfrac{1}{2x}$

The only new twist is that instead of multiplying by a monomial to get a common denominator, you'll sometimes have to multiply by a polynomial. Let's do the thing where I work an example problem and you work another along with me. I'll do the problem $\dfrac{3}{x} + \dfrac{5}{x+1}$ while you do $\dfrac{7}{x} - \dfrac{3}{x-2}$.

My first step is going to be to decide what to multiply each of my fractions by in order to give them a common denominator: x and $(x+1)$ don't share any common factors, so I'm just going to multiply $\dfrac{3}{x}$ by $\dfrac{x+1}{x+1}$ and $\dfrac{5}{x+1}$ by $\dfrac{x}{x}$. Here's what I get:

$$\left(\dfrac{3}{x}\right)\left(\dfrac{x+1}{x+1}\right) + \left(\dfrac{5}{x+1}\right)\left(\dfrac{x}{x}\right) = \dfrac{3x+3}{x(x+1)} + \dfrac{5x}{x(x+1)}$$

(Notice that I've kept the denominator of my fractions in factored form. I like to do that just in case any common factors show up in the numerator later on that I can simplify out.)

21. Decide what to multiply the fractions in your problem, $\dfrac{7}{x} - \dfrac{3}{x-2}$, by and then do so.

Now the two fractions that I'm working with have a common denominator, $x(x+1)$, so they can be added, like this:

$$\frac{3x+3}{x(x+1)} + \frac{5x}{x(x+1)} = \frac{8x+3}{x(x+1)}$$

22. Take a similar step in your problem, remembering that you're subtracting rather than adding.

Actually, we're both done at this point. Neither of our fractions can be simplified, which we can clearly see because we've left our denominators in factored form. Some people do like to multiply out the denominators at this point. You can decide whether or not you want to do so in general, but I'll ask you to do it for this problem just so you keep in mind that your answers may look slightly different from your math partners' and still be equivalent. For my fraction that last step just looks like this:

$$\frac{8x+3}{x(x+1)} = \frac{8x+3}{x^2+x}$$

23. Take that last, usually optional, step for your problem.

And that's really all there is to it, though of course some problems are more complicated than others. Just so you know, I usually don't write out my work quite as elaborately as I did in that example. My actual work would probably look like this:

$$\frac{3}{x} + \frac{5}{x+1} = \frac{3x+3}{x(x+1)} + \frac{5x}{x(x+1)} = \frac{8x+3}{x(x+1)}$$

Do the following addition and subtraction problems and, when possible, simplify the resulting fractions:

24. $\dfrac{5}{x} + \dfrac{2}{x+7}$

25. $\dfrac{3}{x+3} - \dfrac{1}{2x}$

26. $\dfrac{1}{x-2} - \dfrac{1}{x+3}$

27. $\dfrac{x}{x-2} - \dfrac{6}{x^2-x-2}$

(In order to find a common denominator, try factoring the numerator of that second fraction. This time you'll see why it's handy to keep your common denominator in factored form.)

28. $\dfrac{10x}{x^2-x-6} - \dfrac{6}{x-3}$

29. $\dfrac{x^2-8}{x^2+6x+8} + \dfrac{2}{x+2}$

30. $\dfrac{5}{x-3} - \dfrac{15}{x^2-3x}$

31. $\dfrac{1}{x} + \dfrac{1}{x+1} + \dfrac{1}{x-1}$

(When you're adding three fractions, of course you need to find a common denominator for all three.)

32. $\dfrac{x}{x-y} + \dfrac{x}{x+y}$

33. $\dfrac{x-y}{y} - \dfrac{x-y}{x}$

34. $x - \dfrac{1}{x}$

35. $x - \dfrac{x}{y}$

(Remember what the denominator
of x is if you express it as a fraction —
it's the same as the denominator of 2
if you express it as a fraction, as in
"two wholes.")

36. $x + \dfrac{x^2}{5}$

37. $7x - \dfrac{x}{2}$

38. $3 + \dfrac{x-1}{x}$

39. $5 - \dfrac{2}{x+6}$

40. $x - \dfrac{3x}{x+7}$

41. $2x^2 + \dfrac{x^2-5}{4}$

42. $x + 5 + \dfrac{1}{x-5}$ (You can either create two separate fractions out of x and 5 or you can
treat (x + 5) as a single unit and create a single fraction out of it. Either
way gives you the same answer, but I think the second one is easier.)

43. $x + y - \dfrac{2xy}{x+y}$

It's also possible to add or subtract "in reverse" — that is, to create two or more fractions out of a single fraction. And sometimes one of those fractions can be simplified to an integer. For example:

$$\dfrac{3x+5}{3x+2} = \dfrac{3x+2}{3x+2} + \dfrac{3}{3x+2} = 1 + \dfrac{3}{3x+2}$$

44. Write $\dfrac{x+y}{9}$ as the sum of two fractions.

45. Write $\dfrac{3x-5}{x-2}$ as the difference between two fractions.

46. Write $\dfrac{3x+5}{x}$ as the sum of an integer and a fraction.

47. Write $\dfrac{x+y+9}{x+y}$ as the sum of an integer and a fraction.

48. Write $\dfrac{7y-7}{7y+1}$ as the difference between an integer and a fraction.

49. Write $\dfrac{2x^2-2x-5}{2x+5}$ as the difference between a fraction and an integer.

50. Write a *Note to Self* about *adding and subtracting fractions with polynomials.* Explain how to find a common denominator. Make sure that you include at least one example that requires you to simplify the result of your addition or subtraction.

REVIEW

1. Graph the equation $y = 3^{-x}$.

2. Solve the equation $x^2 - x + 1 = x + 64$.

3. Write equations for the set of simultaneous inequalities in this graph:

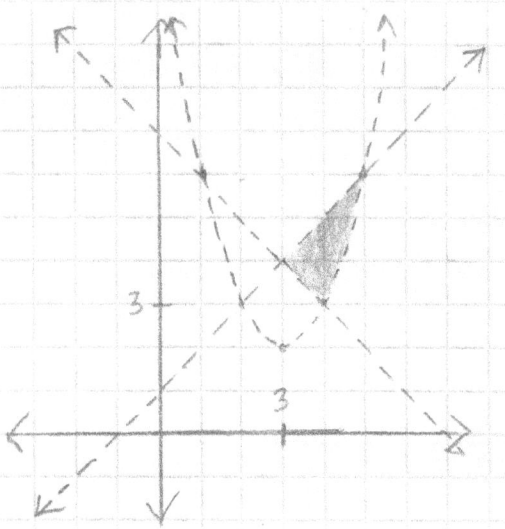

4. Ada, Aaron, and Otis are working together to produce an accurate copy of a painting by Vermeer. If Ada can produce such a copy on her own in 3 months, Aaron can do it in 4 months, and Otis in 5, how long does it take them as a group? (Express your answer in whole months, whole weeks, and approximate number of days.)

Write equations to go with the following tables and state whether their graphs would be straight lines, parabolas, hyperbolas, or exponential curves. (These are challenging ones!)

5.

x	-5	-3	-1	1	3	5
y	4	3	2	1	0	-1

6.

x	-6	-4	0	2	3	6	10
y	-6	-12	12	6	4	3	2

Simplify the following expressions:

7. $\dfrac{(3.5 \cdot 10^{-5})(2.2 \cdot 10^{10})}{1.1 \cdot 10^{-3}}$

8. $(\sqrt{2m^5})(\sqrt{12m^3})$

9. When an object travels in a circle (say it's a mace being swung around the head of a knight, or a unicorn on a carousel, or a car driving around a round-about — it doesn't matter), it's being pulled toward the center of the circle by something that physicists call *centripetal force*. (The actual definition of centripetal force is more complicated than that, but this will do for now.) The formula for centripetal force is:

$$F = \frac{mv^2}{r}$$

... where m is the mass of the object (mace, unicorn, or car), v is the velocity of the object, and r is the radius of the circle. (We won't worry about units for now, either.)

You know about four types of variation: linear variation, varying with the square, inverse variation, and exponential variation (the kind that goes with an exponential function, of course). So...

In this formula, how does the force vary with the mass of the object?
How does the force vary with the velocity of the object?
How does the force vary with the radius of the circle?

10. Here are two more puzzles based on Norman Willis's.

One box contains tiny gold models of fire ants and the other two boxes contain actual fire ants. Here are the labels. (Assume the boxes are numbered left to right.)

Do not open Box #3.	At least one of these labels is false.	If Box #1 is not the box to open, then Box #2 is.

If you must open one of the boxes, which will it be and why?

Or how about this set-up?

All of these labels are false.	The label on Box #1 is true.	This is the correct box to open.

11. Salvador Dalí and an anteater went for a 3-hour jaunt around Paris one fine Sunday. They covered 50 kilometers in all. For the first part of the trip, Salvador rode a scooter at 10 kph and the anteater trotted behind. After a while the anteater was very tired, so they boarded the Metro and rode the rest of the way at 30 kph. How far did they scoot/trot and how far did they ride?

3 MULTIPLYING & DIVIDING FRACTIONS WITH POLYNOMIALS

Multiplying and dividing fractions with polynomials works according to exactly the same principles as multiplying and dividing fractions without polynomials. As I hope you'll recall, multiplying and dividing fractions is often much easier than adding or subtracting them.

Here are a few multiplication and division problems to get you warmed up:

1. $\left(\frac{3}{7}\right)\left(\frac{2}{5}\right)$

2. $\left(\frac{1}{2}\right)\left(\frac{2}{3}\right)\left(\frac{3}{4}\right)$

3. $\left(\frac{x}{2}\right)\left(\frac{x}{3}\right)$

4. $\left(\frac{4}{3}\right)\left(\frac{2}{5}\right)$

5. $\left(\frac{1}{10}\right)(10)$

6. $\left(\frac{x}{6}\right)\left(\frac{x^2}{2}\right)$

There's just one thing I want to go over before I set you loose to do the same thing with polynomials in the fractions. Remember that you ought to simplify a fraction whenever possible, and remember that it's sometimes easier to do the simplifying *before you multiply* (or before you divide — multiplication and division work exactly the same way with fractions). I'll give you an example. Suppose you were going to do this bit of multiplication:

$\left(\frac{60}{7}\right)\left(\frac{9}{220}\right)$

One option would be to multiply **60** by **9** and **7** by **220** to get $\frac{450}{1,540}$. That's perfectly correct. But you need to simplify it — you can tell that $\frac{450}{1,540}$ can be simplified just based on the fact that **450** and **1,540** are both multiples of **10** as well as even numbers — and it is going to be a real beast to simplify. So what you really need to do in order to shortcut the process is to factor the fractions before you multiply. Broken into their prime factors, the two fractions look like this:

$\left(\frac{60}{7}\right)\left(\frac{9}{220}\right) = \left(\frac{2^2 \cdot 3 \cdot 5}{7}\right)\left(\frac{3^2}{2^2 \cdot 5 \cdot 11}\right)$

7. **Check to see that the prime factoring I just did was correct.**

Now you can go ahead and multiply the numerators and the denominators, keeping everything in factored form:

$\left(\frac{60}{7}\right)\left(\frac{9}{220}\right) = \left(\frac{2^2 \cdot 3 \cdot 5}{7}\right)\left(\frac{3^2}{2^2 \cdot 5 \cdot 11}\right)$

$= \frac{2^2 \cdot 3^3 \cdot 5}{2^2 \cdot 5 \cdot 7 \cdot 11}$

... and then several things will "cancel out" of that fraction — meaning that when you divide 2^2 by 2^2 and **5** by **5**, you get **1**:

$$\frac{2^2 \cdot 3^3 \cdot 5}{2^2 \cdot 5 \cdot 7 \cdot 11} = \frac{3^3}{7 \cdot 11}$$

... and that final fraction is pretty easy to multiply out:

$$\frac{3^3}{7 \cdot 11} = \frac{27}{77}$$

Now, that's a pretty labor-intensive process — especially the initial step of prime factorizing **60** and **220** — but for my money it still beats trying to simplify $\frac{450}{1,540}$ down to $\frac{27}{77}$.

Once again, I'm going to work a problem and have you work a similar one alongside me. Here's my problem:

$$\frac{2x + 6}{x^2 - 4} \cdot \frac{x^2 + 3x + 2}{5x^2 + 15x}$$

... and here's yours:

$$\frac{5x + 15}{x^2 + 4x - 5} \cdot \frac{x^2 + 3x - 10}{3x^2 + 9x}$$

As I say, the key is to factor all of the polynomials as completely as possible right out of the gate. So, my first step is going to look like this:

$$\frac{2x + 6}{x^2 - 4} \cdot \frac{x^2 + 3x + 2}{5x^2 + 15x}$$
$$= \frac{2(x + 3)}{(x - 2)(x + 2)} \cdot \frac{(x + 2)(x + 1)}{5x(x + 3)}$$

8. **Take the same step by factoring all of the polynomials in your problem as completely as possible.**

From here the rest is honestly pretty simple. I'm going to do the multiplication, noticing that, for my problem, the (**x + 3**) and the (**x + 2**) are going to "cancel out." I get this:

$$\frac{2(x + 3)}{(x - 2)(x + 2)} \cdot \frac{(x + 2)(x + 1)}{5x(x + 3)} = \frac{2(x + 1)}{5x(x - 2)}$$

9. **Go ahead and do the multiplication in your problem, "canceling out" appropriately.**

Just as we did in the last lesson's addition and subtraction, it's perfectly acceptable to leave our problems in the forms that they're in now, because we can see that they are as simplified as they can be. But at least this once, we'll multiply them out. In my case, that looks like this:

$$\frac{2(x + 1)}{5x(x - 2)} = \frac{2x + 2}{5x^2 - 10x}$$

10. **Take that last, usually optional, step for your problem.**

That's essentially how it works, though as I said in the last lesson, some problems are more complicated than others. But if you just remember to turn division problems into multiplication problems by using the reciprocal of the divisor ("flipping the second fraction over," to put it less formally) and to factor polynomials whenever possible, you'll do just fine.

11. $\dfrac{x^2 - x}{x^2} \cdot \dfrac{x}{x^3 - x}$

12. $\dfrac{2x^2 + 6x}{x} \cdot \dfrac{x^4}{4x^2 + 12}$

13. $\dfrac{\dfrac{x^3}{x^2 - x}}{\dfrac{x^4}{x^3 + x^2}}$

14. $\dfrac{x^2 + x - 2}{x^2} \cdot \dfrac{x^2 + x}{x^2 - 1}$

15. $\dfrac{2x + 6}{x} \cdot \dfrac{5}{x^2 + 6x}$

16. $\dfrac{2x^2 - 2xy}{y + 2} \cdot \dfrac{y^2 + 2y}{2x}$

17. $\dfrac{\dfrac{1}{4x^2 - 9}}{\dfrac{10}{2xy + 3y}}$

18. $\dfrac{6x^3 + 3x^2}{x} \cdot \dfrac{2x - 1}{3x}$

19. $\dfrac{\dfrac{x^2 - 3x - 10}{x - 1}}{\dfrac{x^2 + x - 2}{x^2 - 2x + 1}}$

20. $\dfrac{4x^2 + 8x - 12}{y} \cdot \dfrac{10y}{x^2 + 6x + 9}$

21. $2x^2 - 2y^2 \cdot \dfrac{1}{2x^2 + 4xy + 2y^2}$

22. $\dfrac{x^2 + 2x - 8}{x^2 + 10x + 25} \cdot \dfrac{x^3 + 5x}{x^2 - 4}$

23. $\dfrac{\dfrac{x^2 + x - 30}{3}}{x^2 - 4x - 5}$

24. $\dfrac{2x^2 - 9x - 5}{x^2 - 5x} \cdot \dfrac{x}{2x^2 + 11x + 5}$

25. $\dfrac{\dfrac{x^2 - 25}{3x + 6}}{\dfrac{x^2 + 10x + 25}{9x^2 + 12x - 12}}$

I'd like you to try one more variety of fraction multiplication. Consider the following multiplication problem:

$$\left(x + \dfrac{3}{x}\right)\left(x + \dfrac{5}{x}\right)$$

There are two ways to go about doing this problem, and I'll ask you to try both of them.

26. The first way is to use the FOIL method that you mastered in Chapter 2. (For that matter, you could use the box method if you like, but I've generally found that people prefer FOIL for two-by-two multiplication.) In this case, you'll end up with something that's not quite in proper polynomial form — the last term will be a fraction over x^2 — and that's fine; just leave it that way. The two middle terms should combine into a single term.

27. Now try that same problem a different way. This time you're going to turn each of the portions in the parentheses into a single fraction and then multiply the two fractions as you've been doing in this chapter. Just look at each part in parentheses as a separate fraction addition problem, find the common denominator and rewrite it as a single fraction, then multiply the two fractions.

28. If you did the last two problems the way I did, for Problem 26 you got the answer $\left(x^2 + 8 + \dfrac{15}{x^2}\right)$, and for Problem 27 you got the answer $\left(\dfrac{x^4 + 8x^2 + 15}{x^2}\right)$.

At first glance those two answers may not appear to be the same. Show that they *are* the same by converting one into the other. (I think it's easier to go from the answer to Problem 27 to the answer to Problem 26, but either way will work.)

Do the following multiplication problems using the method you prefer. Whichever method you choose, also write the answer in the form that it would have if you'd used the other method.

29. $\left(x + \dfrac{1}{5}\right)\left(x + \dfrac{4}{5}\right)$

30. $\left(\dfrac{2x}{3} + \dfrac{1}{4}\right)\left(\dfrac{2x}{3} - \dfrac{1}{4}\right)$

31. $\left(x - \dfrac{3}{x}\right)\left(x - \dfrac{10}{x}\right)$

32. $\left(x - \dfrac{3}{5}\right)\left(4x + \dfrac{1}{2}\right)$

33. Before you take a look at complex fractions, write a *Note to Self* about *multiplying and dividing fractions with polynomials*. It should include examples of multiplication and division like the ones up to Problem 25 and also one example of multiplication like the ones from Problems 29 through 32.

Although you may rightly consider many of the fractions that you've dealt with in your mathematical career to be complex, *complex fractions* actually have a specific definition. Here's a complex fraction:

$$\dfrac{1 + \dfrac{2}{5}}{\dfrac{x}{3} - \dfrac{1}{2}}$$

A complex fraction is a fraction that has one or more fractions in the numerator or denominator. Not too surprisingly, a *simple fraction* is a fraction that does not have any fractions in its numerator or denominator.

Any complex fraction can be expressed as a simple fraction, and the process of changing it from a complex to a simple fraction is — perhaps somewhat ironically — not even particularly complex given all the things you know.

Step 1 is to find common denominators for the fractions in the numerator and denominator and convert them into single fractions, like so:

$$\frac{1 + \frac{2}{5}}{\frac{x}{3} - \frac{1}{2}} = \frac{\frac{5}{5} + \frac{2}{5}}{\frac{2x}{6} - \frac{3}{6}} = \frac{\frac{3}{5}}{\frac{2x - 3}{6}}$$

... and now you have a division problem exactly like the ones you've dealt with throughout this lesson:

$$\frac{\frac{3}{5}}{\frac{2x - 3}{6}} = \left(\frac{3}{5}\right)\left(\frac{6}{2x - 3}\right) = \frac{18}{10x - 15}$$

The only thing I need to warn you about is that you should *always* simplify the resulting fraction. And, as you've learned, simplifying fractions sometimes involves factoring polynomials.

Change the following complex fractions into simple fractions in their most simplified forms:

34. $\dfrac{\frac{5}{3} - 1}{\frac{7}{5} - 1}$

35. $\dfrac{4}{\frac{1}{6} - \frac{1}{8}}$

36. $\dfrac{\frac{3}{5}}{2 + \frac{3}{5}}$

37. $\dfrac{x + \frac{3}{5}}{x - \frac{2}{5}}$

38. $\dfrac{\frac{x}{6} + 1}{\frac{x^2}{36} - 1}$

39. $\dfrac{\frac{x}{5} - \frac{x}{6}}{\frac{x}{2} - \frac{x}{3}}$

40. $\dfrac{\frac{x}{9} + 1}{1 + \frac{9}{x}}$

41. $\dfrac{x^2 - 25}{\frac{1}{5} - \frac{1}{x}}$

42. $\dfrac{x - 1 - \frac{6}{x}}{1 - \frac{3}{x}}$

43. $\dfrac{\frac{1}{x} - \frac{3}{x^2} - \frac{10}{x^3}}{\frac{1}{x} + \frac{1}{x^2} - \frac{2}{x^3}}$

44. $\dfrac{x + y}{\frac{1}{x} + \frac{1}{y}}$

45. $\dfrac{x^2 + 4x - 5}{\dfrac{x - 2 + \dfrac{1}{x}}{\dfrac{1}{5} + \dfrac{1}{x}}}$ (This problem has a complex fraction within a complex fraction, but the steps are still the same to simplify it.)

46. Create a complex fraction for a classmate to simplify. Don't just make one up at random, but craft it so that you know what the simplified version will be and so that at least some portions of it will "cancel out" along the way.

47. Write a **Note to Self** about **complex fractions**. The example you use doesn't need to be as complicated as Problem 45!

REVIEW

Simplify the following expressions:

1. $\dfrac{6x^2 - 2xy}{4x}$

2. $\dfrac{x^2 - 4x - 21}{x^2 + 8x + 15}$

State what value or values of the variables are forbidden in the following expressions:

3. $\dfrac{5}{x(x + 2)}$

4. $\dfrac{x}{(m - n)(m + q)}$

5. $\dfrac{3}{x^2 + 4x - 45}$

Simplify the following expressions:

6. $\dfrac{x + 7}{x - 2} - \dfrac{3x - 5}{x - 2}$

7. $\dfrac{5}{x + 1} + \dfrac{3}{x}$

8. $\dfrac{3x}{x + 3} - \dfrac{5x}{x^2 + 4x + 3}$

9. $7 + \dfrac{3}{2x - 5}$

10. Write $\dfrac{5x - 4}{x}$ as the difference between an integer and a fraction.

11. Write $\dfrac{x^2 + 3x + 2}{x + 7}$ as the sum of a fraction and an integer.

Simplify the following expressions:

12. $\dfrac{x^2 + 2x - 35}{x + 2} \cdot \dfrac{x^2 + x - 2}{x + 7}$

13. $\dfrac{5x^2 - 30x + 45}{2x + 3} \cdot \dfrac{2x^2 - x - 6}{15x - 45}$

14. $\dfrac{\dfrac{x^2 - 7x - 18}{6x - 3}}{\dfrac{2x^2 - 22x + 36}{3x - 15}}$

15. $\left(x + \dfrac{4}{x}\right)\left(x - \dfrac{5}{x}\right)$

16. $\dfrac{\dfrac{x}{4} + 1}{\dfrac{x^2}{16} + 1}$

17. Chuckles's swimming pool is three times as long as it is wide. A decorative, slip-proof border, two feet wide, surrounds the pool. The area of the whole affair, including the border, is 171 square feet. What are the dimensions of the pool itself? (Solving this involves factoring a mean-looking polynomial, but it turns out not to be too hard.)

18. Here's quite a different sort of puzzle:

Punctuate (and capitalize, where appropriate) the following string of words in order to make a series of intelligible statements:

that that is is that that is not is not is not that it it is

(There is more than one possible solution.)

5

ADVANCED FACTORING & THE QUADRATIC FORMULA

1 COMPLETING THE SQUARE

At this point, you can solve quite a few quadratic equations. For instance, try this one:

1. **Find the solutions to the equation $x^2 + 3x - 28 = 0$.**

But there are still some quadratic equations that you can't solve.

2. **Explain why you can't use the techniques you've learned so far to solve the equation $x^2 - 8x + 11 = 0$.**

In this lesson you'll develop a new technique that will allow you to solve a lot of currently unsolvable quadratic equations. (And by the end of the next lesson you'll be able to solve literally any quadratic equation.) In order to make sense of this new technique, there are a couple of topics that we have to cover first.

Suppose you were solving the equation $x^2 = 16$. One widely accepted way to write your solutions would look like this:

$x^2 = 16$
$x = \pm 4$

The equation $x = \pm 4$ actually stands for two separate equations. The "$\pm$" is read "plus or minus," so $x = \pm 4$ is read "x equals plus or minus **four**." (Since $x = \pm 4$ represents $x = 4$ *and* $x = -4$, I'm usually going to write as if it's plural — "$x = \pm 4$ *are* the solutions to $x^2 = 16$," not "$x = \pm 4$ *is* the solution" — which may strike your eye in a funny way while you're reading it, but I hope you'll agree with me that it actually makes sense.)

The $\pm$ sign can also be used as part of an expression, such as 5 ± 3, which is equal to either **2** or **8**. If 5 ± 3 were the two solutions to an equation, you could write them in a single equation in one of two ways: either $(x = 5 \pm 3)$ or $(x = 2, 8)$.

For each of the following equations, tell what two numbers x can be equal to:

3. $x = 10 \pm 2$

4. $x = 2 \pm 10$

5. $x = -5 \pm 5$

Okay, that's how the $\pm$ sign works. The next thing we need to do is review how to simplify square roots. Remember that if you have a number that is *not* a perfect square under a radical sign, you should look for factors of that number that *are* perfect squares in order to simplify it. Here, for instance, are the steps for simplifying $\sqrt{18}$:

$$\sqrt{18} = \sqrt{9 \cdot 2} = 3\sqrt{2}$$

Simplify the following expressions:

6. $\sqrt{50}$

7. $\sqrt{48}$

8. $\sqrt{28}$

Now I'd like to look at expressions that combine square roots with the $\pm$ sign, such as $4 \pm \sqrt{5}$. An expression like $4 \pm \sqrt{5}$ refers to two numbers, just the way that 10 ± 2 does. In the case of $4 \pm \sqrt{5}$, however, those two numbers are irrational. They can't be expressed as ratios and they can't be expressed any more simply than $4 \pm \sqrt{5}$. (Notice that, in this case, $\sqrt{5}$ itself is fully simplified. An expression like $4 \pm \sqrt{50}$ could be rewritten as $4 \pm 5\sqrt{2}$, but that's as simple as that one gets.)

If you like, you can estimate the rational value of two numbers like $4 \pm \sqrt{5}$. Remember that the way to estimate the value of a square root is to ask which perfect squares it lies between: **5** lies between **4** and **9**, therefore $\sqrt{5}$ lies between **2** and **3**; **5** is closer to **4** than **9**, so $\sqrt{5}$ is probably pretty close to **2**... let's call it, say, **2.2**. (Checking on the calculator, I found that it's **2.23606...**, so I actually didn't do too badly.) That means that $4 \pm \sqrt{5}$ is roughly either **1.8** or **6.2**.

Estimate the values of each of the following expressions. (Don't use a calculator — and realize that the other math students with whom you check your answers may have made slightly different estimates, which is, of course, just fine — only re-check your work if you have radically different answers.)

9. $7 \pm \sqrt{7}$

10. $10 \pm \sqrt{15}$

11. $4 \pm 2\sqrt{30}$

12. $-6 \pm 2\sqrt{37}$

I won't be asking you to make those sorts of estimates in this lesson again. You can be satisfied with an expression like $7 + \sqrt{7}$ and leave it in that form.

So, the reason that I wanted to cover expressions like $4 \pm \sqrt{5}$ before going on to help you discover the new technique for solving quadratic equations is that the solutions to some quadratic equations have forms like $4 \pm \sqrt{5}$. Remember that $4 \pm \sqrt{5}$ represents two perfectly legitimate numbers, even if they're irrational ones, and that there's no reason why they can't be the solutions to equations.

13. **In fact, $4 \pm \sqrt{5}$ happen to be the solutions to the equation from Problem 2, (x^2 - 8x + 11 = 0), which, you'll remember, you can't solve yet. I am going to ask you to check that those solutions really do make the equation (x^2 - 8x + 11 = 0) true. Honestly, this is probably the hardest thing you'll have to do in this lesson, but I do want you to do it at least once just so you really see that solutions like $4 \pm \sqrt{5}$ can make a quadratic equation true. So go ahead and plug $4 + \sqrt{5}$ and $4 - \sqrt{5}$ into**

($x^2 - 8x + 11 = 0$) to make sure that they both work. The tricky part is that when you plug them into the x^2 part, you'll need to use FOIL. Work carefully and get help from your math partners or teachers if you need to.

Okay, I hope you can now agree with me that it's possible for numbers like $4 \pm \sqrt{5}$ to be the solutions to quadratic equations. Now we're going to build to the new technique by looking first at some pretty simple quadratic equations.

Solve the following quadratic equations by taking the square roots of each side of the equations. Remember that each equation will probably have two answers and simplify the square roots whenever possible.

14. $x^2 = 13$

15. $x^2 = 72$

16. $x^2 = 300$

17. $x^2 = -16$

I threw that last one in as a reminder that not all quadratics have solutions. Well, I should be more precise: not all quadratics have *real* solutions. The square roots of negative numbers are imaginary numbers, so the solutions to an equation like $x^2 = -16$ are imaginary. However, since imaginary numbers go beyond what we're going to study in this textbook, we'll just say that an equation like $x^2 = -16$ has no real solutions (or no non-imaginary solutions, if you prefer to put it that way).

Other than Problem 17, that last little set of problems was pretty easy. The reason it was easy was that the left-hand sides of the equations were all perfect squares, so all you had to do was take the square root of each side. Well, many other quadratics beyond those as straightforward as $x^2 = 16$ or even $x^2 = 72$ or $x^2 = 35$ can be solved by taking the square root of both sides of the equation. Consider this equation:

$x^2 - 6x + 9 = 16$

18. Rewrite the equation ($x^2 - 6x + 9 = 16$), factoring the left-hand side.

19. As you may have realized when you first looked at ($x^2 - 6x + 9 = 16$), ($x^2 - 6x + 9$) is a perfect square. Take the next step in solving the equation by taking the square root of each side. Don't get confused about the square root of $(x - 3)^2$; it really is as simple as it seems. Also, don't forget the $\pm$ sign on the right-hand side.

20. Finish solving the equation by adding 3 to each side. On the right-hand side, you should now have an expression like the ones in Problems 3 through 5. (Actually, you may have decided to write it as "$\pm 4 + 3$." There's nothing wrong with that — either way, you'll get the same results. Just know that "3 ± 4" is the more common way of writing it.) Check those solutions by plugging them into the original equation to make sure you're correct.

Solve the following quadratics and check your solutions in the original equations:

21. $x^2 + 8x + 16 = 36$

22. $x^2 - 10x + 25 = 25$

23. $x^2 + 14x + 49 = 1$

Now, not all quadratics are going to be as well behaved as those. In some cases, even when the polynomial on the left-hand side of the equation is a perfect square, the right-hand side of the equation won't be.

24. **Use the same steps to solve ($x^2 - 6x + 9 = 5$). This time you should end up with an answer that looks like the expressions in Problems 9 through 12.**

In general, as you know, I'm a big believer in checking your work. And as I mentioned in *Jousting Armadillos*, one of the great things about single-variable equations is that you really can check your solutions. Do they make the original equation true? You don't need to ask a classmate or a teacher whether they got the same answer and you don't need to review the steps that you took in solving the equation — you can easily check for yourself whether you're correct. That being said, checking the kind of answer that you just got for Problem 24 is, in my view, generally too much trouble to be worth it. If you're doing a problem like that one on a test, take the time to check the answer. If you become a structural engineer and you're designing a nuclear power plant and you need to solve an equation like that, please, *please* check your answer. But, unless I specify otherwise (and I'll only do so for a good reason), in this math book you can check solutions like $3 \pm \sqrt{5}$ with your math partners. If you guys get different answers, then go back over your work together.

Solve the following quadratics, simplifying square roots when possible:

25. $x^2 + 16x + 64 = 11$

26. $x^2 - 8x + 16 = 6$

27. $x^2 + 20x + 100 = 8$

28. $x^2 - 18x + 81 = -7$

Once again, I put that last one in as a reminder that not all quadratics have real solutions.

All right, now let's look again at the equation from the beginning of the lesson, the one that I said you didn't have the tools to solve yet:

$x^2 - 8x + 11 = 0$

29. **Compare the left-hand side of the equation ($x^2 - 8x + 11 = 0$) to the left-hand sides of the equations in Problems 21 through 28. What makes the left-hand side of ($x^2 - 8x + 11 = 0$) different from those other equations?**

The expression $(x^2 - 8x + 11)$ is impossible to factor. But now that it is part of the equation $(x^2 - 8x + 11 = 0)$, the whole game has changed. The expression $(x^2 - 8x + 11)$ may be impossible to factor, but you know what expression is *easy* to factor? The expression $(x^2 - 8x + 16)$. You factored it in Problem 26. It's easy to factor because it's a perfect square. Wouldn't it makes things easier if, instead of $(x^2 - 8x + 11)$, the left-hand side of the equation we were dealing with was $(x^2 - 8x + 16)$? But wait a minute…

30. **Because $(x^2 - 8x + 11 = 0)$ is an equation, you *can* change the left-hand side to $(x^2 - 8x + 16)$: all you have to do is make sure you add the same thing to both sides of the equation. Go ahead and add the necessary amount to each side of the equation, then solve it exactly the way you solved Problems 25 through 28. You ought to get the same solutions that you checked back in Problem 13.**

The thing you just figured out is the new technique for this lesson. It's called (big surprise here, given the title of the lesson) *completing the square*. The name makes a lot of sense, because what you just did in Problem 30 was to turn the left-hand side of $(x^2 - 8x + 11 = 0)$ into a perfect square: you *completed* the square.

Completing the square will work for a lot of quadratic equations, including some of the ones that you solved by factoring in Chapter 3.

31. **Solve the equation $(x^2 - 6x - 16 = 0)$ by factoring.**

32. **Solve the equation $(x^2 - 6x - 16 = 0)$ by completing the square.**

In the case of an equation like $(x^2 - 6x - 16 = 0)$, it's entirely up to you which method you prefer to use to solve it. The real power of completing the square is that it allows you to solve certain equations, such as $(x^2 - 8x + 11 = 0)$, that you can't solve by factoring.

33. **Before I give you another set of quadratic equations to solve, it might be helpful for you to think for a moment about how to go about completing the square. Here are the beginnings of a few expressions. Figure out what you would need to add to each one in order to make it a perfect square:**

    $x^2 + 4x$ …
    $x^2 + 10x$ …
    $x^2 - 6x$ …
    $x^2 - 10x$ …
    $x^2 + 16x$ …

    How do you choose what number to use in order to make an expression into a perfect square?

    Solve the following equations using whatever methods you choose. Simplify the answers where possible.

34. $x^2 + 8x + 10 = 0$ 35. $x^2 - 10x + 18 = 0$

36. $x^2 + 18x + 49 = 0$ 37. $x^2 + 10x - 15 = 0$

38. $x^2 + 12x = -32$

39. $x^2 - 4x = -4$

40. $x^2 - 16x + 50 = 0$

41. $x^2 = 14x - 40$

42. $x^2 - 20x - 100 = 0$

43. $x^2 + 6x + 13 = 0$

44. $x^2 = 22x$

This time I tried to mix things up a little bit by making the second-to-last problem instead of the last one be the one with no real solutions. If you try to complete the square in order to solve ($x^2 + 6x + 13 = 0$), you end up trying to take the square root of a negative number.

45. **Write three quadratic equations in standard form that you're sure have no real solutions. Your quadratics should include three-term polynomials.**

In general, completing the square is a difficult technique to use on quadratics of the form ($ax^2 + bx + c = 0$) if **b** is an odd number. For example, if you tried to solve ($x^2 + 5x + 5 = 0$) by completing the square, in order to figure out what number to use, you'd need to take half of **5** and square it, which is **6.25**. That means you'd have to add **1.25** to each side of the equation to get:

$x^2 + 5x + 6.25 = 1.25$

Ultimately you'd get the solution pair $-2.5 \pm \sqrt{1.25}$. Yuck. I don't think there's anything technically wrong with that, but in general I'd only try to complete the square if your middle-term coefficient is even.

Completing the square can be used to solve some quadratic equations with first-term coefficients other than one. It will only work when the first term is a perfect square, and it won't always work then. But I'll give you three problems to try where it does work. These are a little trickier than the previous problems, because it's harder to figure out what number you need in order to complete the square. If you're having trouble, my advice is to draw a box like the ones we've used so often before. You'll also notice that your answers are fractions with square roots in them — I definitely won't ask you to check them in the original equations.

Solve the following quadratic equations by completing the square:

46. $9x^2 + 24x + 10 = 0$

47. $16x^2 + 16x - 6 = 0$

48. $4x^2 + 12x - 2 = -6$

49. **Write a *Note to Self* that explains the process of *solving quadratic equations by completing the square*. You can pick one from Problems 34 – 44 or create your own, of course.**

There is one other situation in which completing the square is quite useful, and that's for changing two-variable equations like ($y = x^2 + 2x - 8$) into vertex form (which I told you in Chapter 3 that I'd teach you how to do). It's really pretty simple. So let's figure out how to change ($y = x^2 + 2x - 8$) into vertex form right now.

50. What would you choose to finish this equation:

$y = x^2 + 2x$ ...

... so that the right-hand side would be a perfect square?

51. Explain why ($y = x^2 + 2x - 8$) is equivalent to ($y = x^2 + 2x + 1 - 9$).

52. Take the new equation, ($y = x^2 + 2x + 1 - 9$), and partially factor the right-hand side: that is to say, take the part that is a perfect square and rewrite it as a binomial squared. When you've done this, you should have something that's recognizable as vertex form.

53. Voila. That's all there is to it. Based on the work you just did, where's the vertex of the graph of ($y = x^2 + 2x - 8$)?

Use your factoring skills from Chapter 3 and the technique you just learned to find the x-intercepts and the vertices of graphs of the following equations. (One handy way to check your work is to make sure that the x-coordinate of the vertex is halfway between the x-intercepts.)

54. $y = x^2 - 4x - 12$

55. $y = x^2 + 4x + 3$

56. $y = x^2 - 6x - 27$

57. Notice that, for Problems 54 through 56, the y-coordinates of the vertices are all negative. This shouldn't surprise you, because these are all upward-opening parabolas that cross the x-axis. Assuming that the parabola with the equation ($y = -x^2 + 6x + 16$) crosses the x-axis, what would you expect to be true about the y-coordinate of its vertex and why?

58. You already know how to find the x-intercepts of ($y = -x^2 + 6x + 16$). Just factor -1 out of the polynomial first and then factor the polynomial. Go ahead and do so, and name the x-intercepts.

Finding the vertex form of ($y = -x^2 + 6x + 16$) involves a similar trick. What you have to do is find the *opposite* of the number that you'd usually use to complete the square. I think the best way to explain what I mean is to show you:

$y = -x^2 + 6x + 16$
$y = -x^2 + 6x - 9 + 25$ (-9 is the opposite of the 9 I'd usually need)
$y = -1(x^2 - 6x + 9) + 25$ (now I'm factoring -1 out of part of the equation)
$y = -1(x - 3)^2 + 25$ (and there's vertex form — the vertex is at (3, 25))

I hope that made sense and that therefore you can find the x-intercepts and vertices of the graphs of these equations:

59. $y = -x^2 + 8x - 15$

60. $y = -x^2 + 2x + 48$

61. Write a **Note to Self**, using at least one example, that explains the process of *changing a quadratic equation into vertex form*.

62. Good old Chuckles the Rocket Dog, at the peak of her illustrious career, was fired upward out of her cannon in a way that could be described by this equation:

$h = 128t^2 - 16t$

... where h was her height in feet and t was the time elapsed in seconds. What was the highest point that she reached and how long did she take to reach it? (This is a tricky problem, but it can be solved using the techniques you've learned in this lesson.)

REVIEW

1. Find the x-intercepts of the graph of the following equation:
$y = 3x^2 - 14x - 5$

2. Graph the following equation:
$y = -(x - 3)^2 + 4$

3. Simplify the following expression:
$\sqrt{4x^3} \cdot \sqrt{3x^2} \cdot \sqrt{3x}$

4. Divide $(21x^3 - 64x^2 + 63x - 20)$ by $(7x - 5)$.

5. Solve the following equation:

$$\frac{x + 4}{3} = \frac{4}{x}$$

6. Express $\dfrac{1}{125}$ as a power of 5.

7. Simplify the following expression:

$$-2|-3 \cdot 5| + \frac{2^4}{4}$$

8. Find the slope of a straight line that passes through (-5, -14) and (7, 0).

9. I have five times as many two-pence pieces as five-pence pieces and the total value of my change is 30 pence. What's the mix of coins?

10. Two more puzzles adapted from Norman Willis's work:

You're faced with three boxes, two of which contain odiferous used socks and one that contains cozy new socks. Here are the labels:

No fewer than two of these labels are false.	None of these labels is false.	Open either this box or the one with the label that is true.

Which one do you open and why?

This time there are four lead boxes, three containing radioactive space sludge and one containing non-radioactive space sludge. Here are their labels:

#1: Exactly two of these labels are true.	#2: Open this box.

#3: This is not the box to open.	#4: Box #2 is not the box to open.

Once again, if you're obliged to open a box, which should it be and why?

2 THE QUADRATIC FORMULA

So far you've learned to solve quadratic equations by factoring and by completing the square. Both techniques are very useful, and both have limitations. You can only solve by factoring if the polynomial portion of the quadratic is factorable. You can only solve by completing the square if the first term of the polynomial portion is a perfect square and the second term has an even coefficient. (Actually, as I said in the last lesson, you can solve quadratic equations without those qualities by completing the square — it's just a painful process.) So, when you're faced with a quadratic like ($7x^2 + 14x - 13 = 0$), what do you do? Just the idea of testing whether it's factorable or trying to solve it by completing the square gives me hives.

But there is a way. It's called the Quadratic Formula.

Now, most textbooks would just tell you what the Quadratic Formula was and ask you to use it — which is not particularly hard.

However, I have more faith in you than that, so I am going to take you through the steps of figuring out the Quadratic Formula for yourself. Basically, what you are going to do is complete the square once, in a very tricky way, and then you'll never have to complete the square for a problem like ($7x^2 + 14x - 13 = 0$).

This may very well be the most difficult thing that I ask you to do in any of these three math textbooks, with the possible exception of some of the puzzles in the review sections. But I think you can do it. What you are going to do is use your algebraic skills to prove something, as you have several times before, but this time you'll be proving something profoundly useful rather than just showing that a clever trick will always work.

As I say, I've broken the process down into steps, so that Problems 1 through 13 in this lesson are actually all part of the same problem. You can lay it out however it makes sense to you — as long as it's neat and orderly — in your notebook. I would definitely recommend working on this in class, where you can work with your math partners and potentially ask your teacher for help.

All right, when you're ready, roll up your sleeves and let's get to work…

1. **Since you want to prove something for all possible quadratic equations, the first thing to do is to set up a completely generic quadratic equation in standard form, like so: ($ax^2 + bx + c = 0$). Now, subtract c from each side.**

As I said, what you're actually going to be doing is completing the square using this generic quadratic. I think the best way for me to help you through this is for me to solve three non-generic quadratics by completing the square. Then you can use those as a pattern for solving your generic quadratic (which will be considerably harder). The quadratics I'll solve will be ($4x^2 + 12x = -5$), ($9x^2 + 30x = -10$), and ($25x^2 + 20x = -7$). Notice that I already made them match your generic ($ax^2 + bx = -c$).

The first step I'd take in completing those squares would be this:

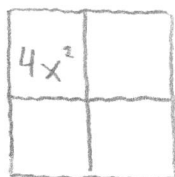

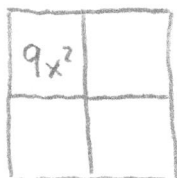

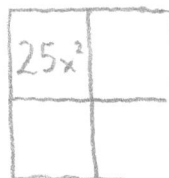

(I left the usual + signs out of my work just so it looks less cluttered.)

2. **Go ahead and take the same step with your generic quadratic.**

Not too bad so far, right? Now it gets a little trickier. The next step for ($4x^2 + 12x = -5$), ($9x^2 + 30x = -25$), and ($25x^2 + 20x = -7$) looks like this:

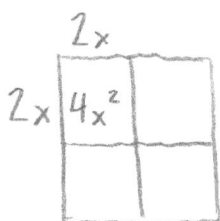

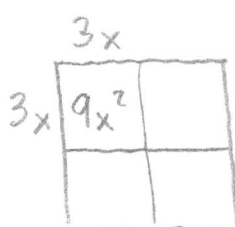

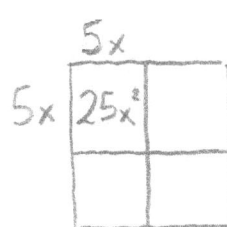

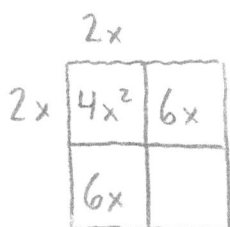

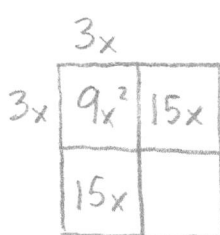

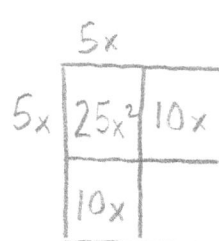

3. **Now you need to make the same move with your generic quadratic, which is more difficult than what I had to do. Remember that the a in your ax² corresponds to the 4 in my 4x², the 9 in my 9x² and the 25 in my 25x². What should you have in place of the 2 in my 2x, the 3 in my 3x and the 5 in my 5x? Once you've decided what goes on the edges of your box, test to see that you really do get ax² when you multiply those two things.**

The next step for ($4x^2 + 12x = -5$), ($9x^2 + 30x = -25$), and ($25x^2 + 20x = -7$) looks like this:

4. **Take the same step for your generic quadratic. Again, the b in your bx corresponds to the 12 in my 12x, the 30 in my 30x, and the 20 in my 20x. So what should go in the upper right and lower left compartments of your box? There are two ways to write the expression you need to write. Both are perfectly fine and both involve a fraction.**

The next step for ($4x^2 + 12x = -5$), ($9x^2 + 30x = -25$), and ($25x^2 + 20x = -7$) is to ask these questions:

"What do I have to multiply **2x** by to get **6x**?"
"What do I have to multiply **3x** by to get **9x**?"
"What do I have to multiply **5x** by to get **10x**?"

5. **Looking at your box, just write down the equivalent question for your generic quadratic.**

Of course, now you're going to have to answer the question. For the three quadratics I'm working on, the questions are pretty easy to answer. For yours, I think this may be the hardest step of the whole process. When I was going through the steps that you're doing, I found that the easiest way for me to think about this problem was to write it out in this kind of form:

(2x)(_____) = 6x

Alternatively, you could write a division problem similar to this one:

$$\frac{6x}{2x} =$$

6. **Answer the question that you wrote in Problem 5. However you decide to tackle it, if you want a hint, you should know that your answer will be a fraction with a square root as part of the denominator.**

When I've answered the questions from Problem 5, the next step for ($4x^2 + 12x = -5$), ($9x^2 + 30x = -25$), and ($25x^2 + 20x = -7$) is to fill in the answers like this:

7. **Fill in the same spots in your work using your answer from Problem 6.**

For ($4x^2 + 12x = -5$), ($9x^2 + 30x = -25$), and ($25x^2 + 20x = -7$), I can now, of course, do this:

8. Take the same step for your generic quadratic. It will be a little trickier in your case than it was in my three equations — but really you're just doing a fraction multiplication problem that involves a square root.

Filling in that lower right compartment was the whole point of everything we've done so far, since it tells us what we need to add to each side of our equations in order to complete the squares. In my case, I'm adding **9**, **25**, and **4**. My next steps, therefore, look like this:

$4x^2 + 12x = -5$
$4x^2 + 12x + 9 = 9 - 5$

$9x^2 + 30x = -25$
$9x^2 + 30x + 25 = 25 - 25$

$25x^2 + 20x = -7$
$25x^2 + 20x + 4 = 4 - 7$

9. Go back to your generic equation ($ax^2 + bx = -c$) and add the appropriate thing to each side. If I were really solving the equations I'm working with, I would simplify the right-hand sides — for instance, I'd write "4" instead of "9 - 5" – but I left them that way to give you a better idea of the form your equation should have at this point.

10. Now you should take a step that I don't need to. As you know, doing algebra with fractions can be kind of difficult. Remember that one of the best ways of dealing with fractions in algebraic equations is to multiply the entire equation by whatever you need to to get rid of the fraction.

For instance, in the equation $x + \frac{x}{m} + \frac{x}{5} = 10$, I'd multiply the entire thing by 5m to

get ($5mx + 5x + mx = 50m$). You should multiply your entire equation by whatever will get rid of the two fractions in it. Just be sure to multiply every term by that same thing.

Now, if we've been doing everything right, the expressions on the left-hand sides of our equations should be perfect squares, and we can factor them. For the three I'm working with, that looks like this:

$4x^2 + 12x = -5$
$4x^2 + 12x + 9 = 9 - 5$
$(2x + 3)^2 = 9 - 5$

$9x^2 + 30x = -25$
$9x^2 + 30x + 25 = 25 - 25$
$(3x + 5)^2 = 25 - 25$

$25x^2 + 20x = -7$
$25x^2 + 20x + 4 = 4 - 7$
$(5x + 2)^2 = 4 - 7$

For your equation, that factoring will be slightly trickier. You may be able to do it straight off, but when I was working on it, I had to make a new box that started like this:

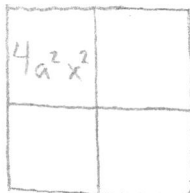

(There's a little hint for you as to whether you're on the right track at this point...)

11. **Whether or not you decide to use a box to do the factoring work, you should factor the perfect square on the left-hand side of your equation as I did for my equations above and write the new version of the equation.**

You're getting very close now. The next step, as you know very well, is to take the square root of each side of the equation. For one of my equations, that would look like this:

$4x^2 + 12x = -5$
$4x^2 + 12x + 9 = 9 - 5$
$(2x + 3)^2 = 9 - 5$
$2x + 3 = \pm\sqrt{9 - 5}$

12. **Take the same step in your work, remembering to include the $\pm$ sign.**

13. **You don't need me to walk you through the last two steps; they're simple old algebra. Just remember a couple of things. First, $\sqrt{x+y}$ is not equal to $\sqrt{x} + \sqrt{y}$, so basically the stuff that you have under the radical sign right now is going to stay there. Second, when you're doing the final division step, you should divide the entire right-hand side by whatever you're dividing the left-hand side by. The right-hand side of the Quadratic Formula is traditionally expressed as one long, rather complicated-looking fraction.**

Okay, sit back for a minute. If you're in my classroom, at any rate, you get to take a break. You have just derived the Quadratic Formula. Trust me when I say that very few people can make that claim.

Just to be sure that you've got it right, check in with a teacher. There's a traditional form to the Quadratic Formula; your version may have a slightly different form — which is fine! — but you'll want to see the traditional one as well.

I'm not necessarily a big proponent of memorizing things, but many students do memorize the Quadratic Formula, and since you may encounter teachers who will expect you to have it memorized, you probably ought to go ahead and do so. It's usually read out loud like this: "**x equals negative b plus or minus the square root of b squared minus 4ac, all over 2a.**" Apparently there's even a tune to sing it to. Perhaps your teacher can teach you the tune — or you can come up with your own if you like.

Now let's take a look at how it's used.

As I've said, the beauty of the Quadratic Formula is that it can be used to solve *any* quadratic equation. That's because it provides a general recipe based on the two coefficients (**a** and **b** in the formula) and the constant (**c** in the formula) of any quadratic equation.

In order to use the Quadratic Formula to solve a quadratic equation, you just have to plug in the values of **a**, **b**, and **c**. As an example, I'll use it to solve the quadratic that I started the lesson with: ($7x^2 + 14x - 13 = 0$). In this case, a = 7, b = 14, and c = -13. So, here's the Quadratic Formula applied:

$$x = \frac{-14 \pm \sqrt{14^2 - (4)(7)(-13)}}{2(7)}$$

$$= \frac{-14 \pm \sqrt{196 - (-364)}}{14}$$

$$= \frac{-14 \pm \sqrt{560}}{14} = \frac{-14 \pm \sqrt{(16)(35)}}{14}$$

$$= \frac{-14 \pm 4\sqrt{35}}{14} = \frac{-7 \pm 2\sqrt{35}}{7}$$

Did I say that the solutions to that quadratic equation would be pretty? I will definitely not make you go back and check those solutions in the original equation. But even though using the Quadratic Formula to solve ($7x^2 + 14x - 13 = 0$) took several steps and I had to be careful along the way, imagine even attempting to solve it by factoring!

Make sure that you look through the steps that I took in solving ($7x^2 + 14x - 13 = 0$) and that you understand each one. In particular, notice that I simplified it as much as possible by simplifying the square root and then, in the last step, simplifying the fraction as much as possible. Simplicity, generally speaking, is best.

Simplify the following expressions as much as possible:

14. $\dfrac{-8 \pm \sqrt{80}}{2}$

15. $\dfrac{10 \pm \sqrt{12}}{12}$

Use the Quadratic Formula to solve the following quadratics:

16. $3x^2 + 8x + 3 = 0$

17. $3x^2 + 5x + 2 = 0$

18. $x^2 - 10x + 5 = 0$

19. $3x^2 - x - 4 = 0$

20. $2x^2 - 2x + 5 = 0$

Notice that the Quadratic Formula doesn't always produce ugly-looking answers. The solutions to Problems 17 and 19 were actually rational numbers.

The Quadratic Formula really can be used to solve *any* quadratic.

21. Solve (x^2 - 1 = 0) by factoring.

22. Solve (x^2 - 1 = 0) using the Quadratic Formula.

23. Solve ($2x^2$ + x - 1 = 0) by factoring.

24. Solve ($2x^2$ + x - 1 = 0) using the Quadratic Formula.

25. Based on your answers to Problems 23 and 24, look back at Problem 17 and decide what the factored form of ($3x^2$ + 5x + 2) must be. Test to see that when you multiply those factors out, you really do get ($3x^2$ + 5x + 2). Do you think it would have been easier to solve Problem 17 using factoring than using the Quadratic Formula? Why or why not?

When I was working on creating quadratic equations for you to solve using the Quadratic Formula, I began to wonder why I never saw quadratic equations in other textbooks where the initial coefficient of the polynomial was negative, such as (**-$2x^2$ + 3x - 1 = 0**). I thought I might give you a few problems like that, but then I figured out why other textbooks don't use them.

26. Solve (-x^2 - 2x + 3 = 0) using the Quadratic Formula.

27. Solve (x^2 + 2x - 3 = 0) using the Quadratic Formula.

28. Multiply the entire equation (-x^2 - 2x + 3 = 0) by -1 and use the result of that multiplication to explain the results of Problems 26 and 27. Based on this, why don't textbooks bother to use quadratics like (-x^2 - 2x + 3 = 0)?

Use the Quadratic Formula to solve the following quadratics. (Just for fun, they're the three that I used as examples when you were deriving the Quadratic Formula.)

29. $4x^2$ + 12x + 5 = 0

30. $9x^2$ + 30x + 25 = 0

31. $25x^2$ + 20x + 7 = 0 (If you're paying close attention while you solve this one, you'll realize that you don't need to bother simplifying the solutions very far.)

32. How many real solutions does the quadratic in Problem 29 have? How about the one in Problem 30? How about the one in Problem 31?

As you probably realized when you did those last problems, there is a particular piece of the Quadratic Equation that tells you how many solutions each of those equations have. It's the part under the square root sign: $(b^2 - 4ac)$. If $(b^2 - 4ac)$ is positive, the quadratic has two real solutions. If $(b^2 - 4ac)$ is equal to zero, the quadratic has one real solution. (After all, the "plus or minus" part has basically disappeared.) If $(b^2 - 4ac)$ is negative, then the quadratic has no real solutions because you can't take the square root of a negative number without dealing with imaginary numbers. (It's still true that the Quadratic Equation can be used to solve any quadratic. After all, I'd argue that "no real solutions" is a kind of solution. Besides that, once you study imaginary numbers, you'll find that the Quadratic Formula will give you the imaginary solutions to a quadratic.)

Because $(b^2 - 4ac)$ allows you to discriminate (in the sense of "tell the difference") between quadratics that have two, one, or no real solutions, $(b^2 - 4ac)$ is known as the *discriminant*.

Use the discriminant to determine how many real solutions each of the following quadratics has:

33. $x^2 + 5x + 5 = 0$

34. $x^2 + 5x + 6 = 0$

35. $x^2 + 5x + 7 = 0$

36. $x^2 - 5x + 6 = 0$

37. $2x^2 + 4x + 4 = 0$

38. $2x^2 + 4x - 4 = 0$

39. $15x^2 + 2x + 1 = 0$

40. $15x^2 + 2x - 1 = 0$

41. $3x^2 - 6x + 3 = 0$

42. Create three quadratics, one with two real solutions, one with one real solution, and one with no real solutions. Have your math partners check to see that the quadratics really do have those numbers of solutions.

Solve the following quadratics using whatever method you like. (The Quadratic Formula can, of course, be used to solve all of them, but you may determine that there are some for which you prefer to use other methods.) In some cases, you'll need to manipulate the equations before you solve them.

43. $2x^2 + 4x - 6 = 0$

44. $7x^2 + x - 2 = 0$

45. $x^2 - 10x + 20 = -5$

46. $x^2 + 4x = -2$

47. $2x^2 - 5 + 2x^2 = -3x$

48. $3x^2 - 5 = 2x$

49. $9x^2 + 6x + 6 = 5$

50. $4x^2 + 6x = 5 - 5x^2$

51. $x^2 + 7x + 13 = 0$

52. $x(4x + 1) = 5$

53. $x^2 = x + 1$

54. Write two *Notes to Self*, one about *the Quadratic Formula* and one specifically about *how to use the discriminant*. (If you prefer, you can write a single Note, but it seems clearer to me to make two separate ones.) You should, of course, include examples of how to use both the Quadratic Formula and the discriminant.

REVIEW

1. Graph the equation $\dfrac{x^2}{5} + \dfrac{y^2}{10} = 1$, estimating the x- and y-intercepts.

Apply the Laws of Exponents appropriately to the following expressions:

2. $(2y^5 - 6y^5)^3$

3. $((m^2 n^3)^4)^5$

4. Simplify the expression $\dfrac{4^{-2} x^5 y^{-3} z^2}{3^{-2} x^{-2} y^{-3} z^5}$ and express your answer without any negative exponents.

5. Find the x-intercepts of the graph of $y = -2x^2 + 16x + 66$.

6. Factor the following expression as a difference of squares and then simplify: $(x + 4)^2 - (x - 2)^2$

Simplify the following expressions:

7. $(x^2 - 20x + 100)\left(\dfrac{x + 3}{x^2 - 7x - 30}\right)$

8. $\sqrt{2}(\sqrt{18} - \sqrt{32})$

9. Bertie and Jeeves took a leisurely motorcar ride out to Brinkley Manor at 15 miles an hour. After an unfortunate incident involving one of Bertie's aunts, a policeman's helmet, and a newt, they returned home to London by the same route at the brisker pace of 25 miles an hour. If the round trip took a total of 8 hours, how far from London is Brinkley Manor?

10. Here's a puzzle from BrainBashers.com:

A number can be described by the following rules:

1. If it is not a multiple of 4, then it is between 60 and 69, inclusive.
2. If it is a multiple of 3, then it is between 50 and 59, inclusive.
3. If it is not a multiple of 6, then it is between 70 and 79, inclusive.

What is the number?

3 FACTORING HIGHER-DEGREE POLYNOMIALS

Up to this point, you've worked mostly with second-degree polynomials and quadratic equations. The last things you'll learn in this book are how to factor *some* higher-degree equations and then how to apply those skills to solving and graphing higher-degree equations. I think you'll find that some of the problems at the beginning of this lesson are quite challenging, but if you stick with them and get help as necessary, it will get substantially easier from about Problem 12 on.

So, it will sometimes be possible to apply the factoring techniques that you've learned already in this book to polynomials of higher than second degree. In particular, the box technique will sometimes work. We'll take a look at a few polynomials that it does work on, starting with this one:

$$x^3 + 2x^2 - 5x + 2$$

If I wanted to try factoring that polynomial into two other polynomials, the first thing I'd do is set up a box like this:

1. **Why do you think I chose a three-compartment by two-compartment box to do my factoring in? (In order to answer this, you might consider what the other alternatives were, keeping in mind that I'm trying to factor ($x^3 + 2x^2 - 5x + 2$) into *two* other polynomials.)**

2. **How do I know that the x^2 belongs in the trinomial and the x belongs in the binomial? (Again, consider what the alternative is — and why it wouldn't work.)**

From here on out, factoring ($x^3 + 2x^2 - 5x + 2$) works a lot like the previous box-technique factoring that you've done. Fortunately there's a pretty limited number of factor pairs to test for **2**, so this one shouldn't be terribly hard. I'll walk you through it.

3. **Copy the box above into your notebook. I've listed the factor pairs for 2 to the right of the box and written the polynomial above the box, the way I think you ought to. Now there's really nothing to do other than start testing those pairs, and since the polynomial doesn't seem to provide any clues as to which pair is correct, I don't see any reason not to do them more or less randomly. I'm going to engineer your first guess, and I'll let you know now that it's going to be wrong, but guess it anyway so I**

can take you through the process. Your guess is that (+ 2) goes above the box on the far right, as part of the second-degree polynomial, and (+ 1) goes beside the box to the lower left as part of the first-degree polynomial. Write this guess down and it will enable you to fill in the top right and lower left compartments of your box, which you should do. (Write lightly because you're going to end up erasing it.)

4. Since you're trying to get $(x^3 + 2x^2 - 5x + 2)$, know that you're aiming to have $(- 5x)$ as one of your two middle terms. Right now you have $2x$ in the upper right compartment, so what has to go in the lower middle compartment? Fill the lower middle compartment in. (Keep writing lightly.)

5. Based on what you put in the lower middle compartment, only one thing can go above the middle row as the middle term of the second-degree polynomial. Fill that thing in.

6. Now you have the information you need to fill in the last compartment. The only question is, when you add the thing you put in that compartment to x^2, do you get $2x^2$? You won't, but you should double-check mentally that you don't. Now you should erase the necessary things so that your box goes back to looking as it did on the first page of this lesson and keep on trying combinations. Now that you see how this works, I don't think it will take you long to find the right combination. Go for it. Once you've found that combination, be sure to write the factored form of $(x^3 + 2x^2 - 5x + 2)$ somewhere near your work.

7. Use similar logic and techniques to factor $(x^3 - 10x^2 + 22x - 7)$ into two other polynomials.

8. Do the same thing for $(x^3 - 2x^2 - 14x + 15)$. (Since 15 isn't prime, there are more possible factor pairs for this one.)

9. Do the same thing for $(2x^3 + 3x^2 - 11x - 6)$. (This one is considerably harder because this time there are even more possible factor combinations — you'll need to list some to the left of your box as well.)

10. This is the hardest bit of factoring that I'll ask you to do in this book. In fact, it's hard enough that I'm going to say you should work on it for a maximum of fifteen minutes or until you are feeling like you'd rather jump out a window than suffer any more of this, and then you should stop. Factor $(x^4 + 7x^3 + 8x^2 + 17x - 3)$ into two trinomials.

11. All right, I promise this is easier than the last one. You learned in Chapter 3 that the sum of squares can't be factored using the techniques you'll learn in this book, but the sum of *cubes* actually can. Factor $(x^3 + y^3)$ as the product of a binomial and a trinomial.

Excellent work. Honestly, you're doing more advanced work than many algebra textbooks would ask you to do at this level. In fact, I'm going to call applying the box technique to higher-degree polynomials "advanced factoring," just so that I have a way to refer to it. Now I'd like to look briefly at the limits of advanced factoring.

I think Problem 10 is genuinely hard — maybe even pushing the limits of advanced factoring already. Maybe you got lucky and figured out the right combination of factors on your first guess, but I hope you can appreciate that it is at least potentially quite complicated. Now suppose you were going to factor this polynomial:

$$x^6 - 3x^5 - 8x^4 + 22x^3 - 19x^2 + 13x - 6$$

I happen to know that it *can* be factored because I made it up by multiplying two other polynomials. I even think it's theoretically possible to factor it using the box technique. But consider what would be involved. First off, in Problem 10 I told you beforehand that it would factor into two trinomials. If I didn't give you the same hint for this polynomial, I think there are at least three possible variations for the initial set-up:

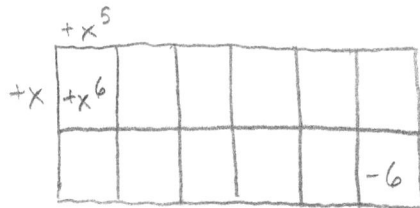

or

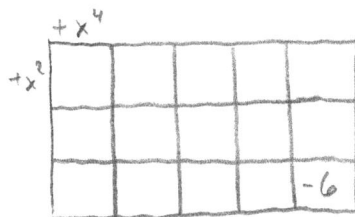

or

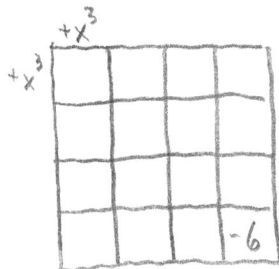

... and those are just the possibilities for the initial set-up. Now you've got to look at the factors of **-6**: **(1)(-6)**, **(-1)(6)**, **(2)(-3)**, and **(-2)(3)**. You'd have to try every one of those combinations in each of the possible set-ups and in each case you'd have to start dealing with all of the possible factor combinations that would crop up for the middle terms. We're already starting to talk about a bewildering number of combinations to test, and **-6** isn't even really a number with a lot of factors, *and* the **x⁶** doesn't even have a coefficient, which, as you know, tends to make factoring roughly twice as difficult. Plus, what if it turned out that I'd got my calculations wrong and ($x^6 - 3x^5 - 8x^4 + 22x^3 - 19x^2 + 13x - 6$) was actually a prime polynomial and you really had to try every single one of those possible combinations without finding one that works?

It makes my head hurt just to think about it.

So here's what I'm really trying to say: I don't expect you to try to employ advanced factoring techniques on every higher-degree polynomial that you encounter. In fact, I won't ask you to use them at all for the rest of the book. I just wanted you to know that it was possible to use the techniques you've learned on quite complex polynomials. So I'll continue to expect you try to factor second-degree polynomials, but I'll only ask you to factor higher-degree polynomials in a couple of specific situations that I'll go over now.

The first is when you have a special case like the ones that you encountered back in Chapter 3, Lesson 5. When a trinomial falls into the pattern where the exponent of the second term is half the exponent of the first term and the third term is a constant, you can easily apply the techniques that you know.

12. Factor (x^6 - $3x^3$ - 10) using the technique of finding two numbers whose product is -10 and sum is -3.

The only other situation where I'll expect you to factor higher-degree polynomials at this point in your career is when you can first factor out a common factor and then use your other factoring techniques. Let's look at an example. Say, for instance, that I asked you to factor ($2x^3$ + $4x^2$ - 30x).

13. Find the greatest common factor of the terms of ($2x^3$ + $4x^2$ - 30x) and factor it out.

14. Now look at the trinomial part of your factored version. It can be factored into two binomials. Go ahead and do so.

Make sure that you always write out the original and the fully factored version of the polynomial. For example, if I were factoring ($2x^4$ + $6x^3$ - $36x^2$), I would first factor $2x^2$ out of all of the terms to get ($2x^2$)(x^2 + 3x - 18). Then I'd factor (x^2 + 3x - 18) into (x + 6)(x - 3). No matter how I went about taking those steps, I would be absolutely sure that one of these two things appeared on my page:

$2x^4$ + $6x^3$ - $36x^2$ = $2x^2$(x^2 + 3x - 18) = $2x^2$(x +6)(x - 3)

... or at least:

$2x^4$ + $6x^3$ - $36x^2$ = $2x^2$(x + 6)(x - 3)

To my mind, the first version is preferable, but the second version is absolutely necessary at minimum. This is not only so that you show your work clearly, but so that you keep in mind what you are actually doing, which is finding expressions that are equivalent to each other.

All right, you are very close to mastering the factoring techniques that you will learn in this book. In a little bit I'll give you a big problem set of higher-degree polynomials to factor, but before that I want to give an overview of what I think you ought to be able to do. First off, I'll lay out the steps of factoring higher-degree polynomials in a tree diagram. This may be helpful to you — it's helpful for me, at any rate. It looks like this:

Factoring Higher-Degree Polynomials

If possible, always factor out the greatest common factor of the terms of the polynomial.

e.g. $2x^3 - 6x^2 + 10x = 2x(x^2 - 3x + 5)$

If you now have a second-degree polynomial to factor...

If you have a polynomial that falls into the special pattern $x^6 + 6x^3 + 5$, factor it appropriately
e.g. $x^6 + 6x^3 + 5 = (x^3 + 5)(x^3 + 1)$
If you have a difference of squares, factor it appropriately.
e.g. $9x^4 - 1 = (3x^2 + 1)(3x^2 - 1)$

If you have a higher degree polynomial that does not fall into the pattern $x^6 + 6x^3 + 5$, use advanced factoring techniques if you choose.

If you are dealing with the difference of squares, factor appropriately.
e.g. $x(x^2 - 16) = x(x + 4)(x - 4)$

If you are dealing with a trinomial with a first-term coefficient of 1, attempt to factor it by finding factors of the third term whose sum is the coefficient of the second term.
e.g. $6x(x^2 - 2x - 15) = 6x(x - 5)(x + 3)$

If you are dealing with a trinomial with a first-term coefficient other than 1, attempt to factor it using the box method.
e.g.
$x(6x^2 - 11x - 10) = x(3x + 2)(2x - 5)$

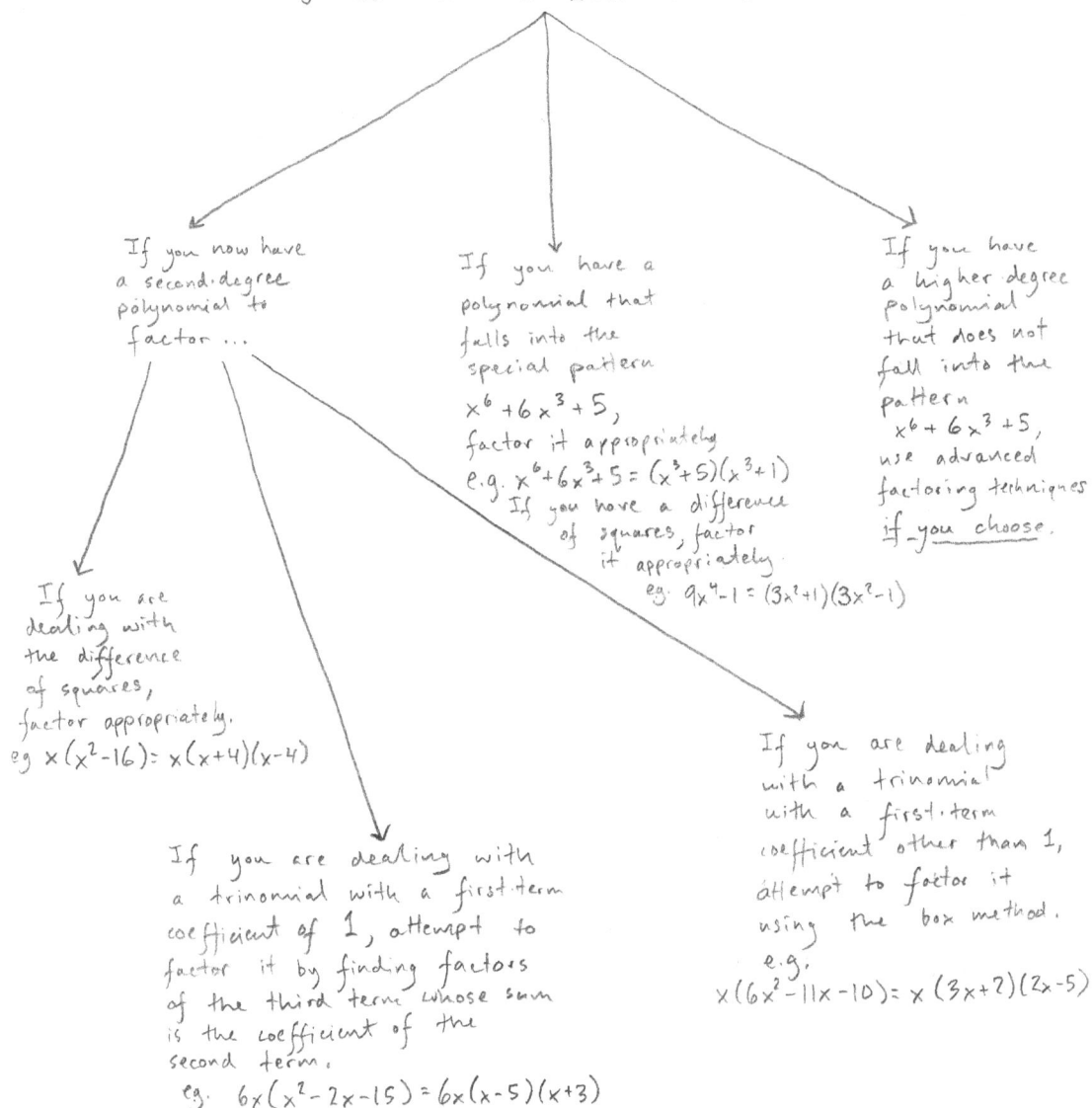

Notice a few things. First off, the examples in this diagram are polynomials with only one variable, but remember that most of the techniques can be used for polynomials with more than one variable.

Second, notice that the diagram suggests that you use advanced factoring techniques *if you choose*. I mean that. You can try them if you want a challenge, but I don't think any reasonable teacher would expect you to use them regularly at this point in your career.

Third, notice that I've written things like, "If possible, factor..." or, "attempt to factor...". Remember that polynomials can be prime. Suppose, for instance, that you're factoring $(2x^3 + 14x^2 + 6x)$. You can factor out **2x** like so:

$(2x^3 + 14x^2 + 6x) = 2x(x^2 + 7x + 3)$

Now you've got a trinomial whose first-term coefficient is **1**, so you should try to factor it using the technique from Chapter 3, Lesson 2. But you'll quickly see that there are no factors of **3** that add up to **7**. So the step of factoring out **2x** is all you can do with ($2x^3 + 14x^2 + 6x$).

Just in case working with diagrams isn't the way you like to go, I'll present the same information (more or less) as a list of the techniques. In each case, you apply the technique only if it's possible. I've given an example of each technique — again using only one-variable polynomials.

1. **Factor out the greatest common factor of the terms of the polynomial.**
   $3x^3 - 6x^2 + 9x = 3x(x^2 - 2x + 3)$

2. **Factor the difference of squares.**
   $9x^4 - 1 = (3x^2 - 1)(3x^2 + 1)$

3. **Factor higher-degree polynomials where the exponent of the middle term is half the exponent of the first term and the third term is a constant.**
   $x^{10} - 5x^5 + 6 = (x^5 - 3)(x^5 - 2)$

4. **Factor second-degree trinomials whose first-term coefficient is 1 by finding factors of the third term whose sum is the coefficient of the middle term.**
   $x^2 - 8x + 7 = (x - 7)(x - 1)$

5. **Factor second-degree trinomials whose first-term coefficient is not 1 by using the box method.**
   $8x^2 + 22x - 6 = (2x + 6)(4x - 1)$

So think about the steps of factoring higher-degree polynomials in any way you like — a diagram like mine, a list of techniques, or anything else that makes sense to you. I'm going to give you a bunch of higher-degree polynomials to practice on.

Factor the following polynomials as completely as possible:

15. $3x^3 - 12x^2 - 63x$

16. $5x^3 - 20x^2 + 20x$

17. $4x^4 - 144$

18. $2x^8 + 4x^4 - 6$

19. $x^4 - 1$ **(This one is actually a little tricky — it can be factored further than you might think.)**

20. $5x^3 - 25x^2 + 20x$

21. $2x^3 - x^2 - 15x$

22. $x^4 - 81$

23. $x^{12} + 6x^{11} + 9x^{10}$

24. $12x^4 - 75x^2$

25. $2x^{10} - 18$

26. $x^4 + 13x^3 + 40x^2$

27. $15x^3 + 72x^2 - 15x$

28. $x^4 - 5x^3 - 9x^2 + 45x$

29. $9x^3 + 63x^2 - 270x$

30. $9x^6 + x^4$

31. $x^3y - 4xy^3$

32. $2x^3 + 5x^2y + 2xy^2$

33. $5x^3y + 15x^3 - 10x^2y - 30x^2$ (Although this polynomial has four terms, it actually functions according to the factoring rules for a trinomial.)

34. $9x^4 + 8x^2y^2 - y^4$

35. It's time for a **Note to Self** about *factoring higher-degree polynomials.* Along with at least one good example, it should probably include the diagram from page 184 or the list from page 185 or your own version of how to keep track of the possibilities for factoring higher-degree polynomials.

REVIEW

For each of the following equations, state whether its graph would be a straight line, a parabola, a hyperbola, an exponential curve, or a non-functional curve. Assume that x is the independent variable, y is the dependent variable, and all other letters are constants.

1. $2y + 5^{x-10} = y$

2. $y = \dfrac{m}{x + 15} - n$

3. $(x - 5)^2 = (y + 2)^2$

4. $\dfrac{x}{1,000m} = \dfrac{y}{n}$

5. Consider the two equations $y^2 = x$ and $y = \sqrt{x}$. Although they are similar, there is a key difference between them. Explain why only one of them can be a function. (It may be helpful to graph them.)

6. Divide $(-4x^2 + 7x - 3)$ by $(-2x^2 - 6x - 10)$.

7. Simplify the following expression:

$(5.72 \cdot 10^{17})(3.04 \cdot 10^{-5})$

8. Linus and the armadillo are playing poker. Linus starts the evening with half again as much money as the armadillo, but over the course of the night he loses half his money to the armadillo. If they're the only two players and the armadillos finishes with $105, how much did each player start with?

9. Here is a classic paradox that you are now equipped to understand.

This is a proof that 2 = 0:

Assume that x = 2: $x = 2$
Multiply both sides by x: $x^2 = 2x$
Subtract 4 from both sides: $x^2 - 4 = 2x - 4$
Factor both sides: $(x + 2)(x - 2) = 2(x - 2)$
Divide both sides by (x - 2): $x + 2 = 2$
Subtract 2 from both sides: $x = 0$

Therefore, 2 = 0!

Where is the flaw in this proof?

4 SINGLE-VARIABLE EQUATIONS OF HIGHER-DEGREE

I mentioned, back when you were learning to solve quadratic equations in Chapter 3, that it's quite possible to have equations that are higher than second degree. It's time to take a look at how to solve some of them — and how to recognize when you can't — now.

Here's a table representing the generic forms of some of those equations in standard form:

Degree	Generic Equation in Standard Form	Name
1	$ax + b = 0$	linear
2	$ax^2 + bx + c = 0$	quadratic
3	$ax^3 + bx^2 + cx + d = 0$	cubic
4	$ax^4 + bx^3 + cx^2 + dx + e = 0$	quartic
5	$ax^5 + bx^4 + cx^3 + dx^2 + ex + f = 0$	quintic

In the table above, **a**, **b**, **c**, **d**, **e**, and **f** can have any value except that **a** can't be zero.

1. **Why can't a be equal to zero in that table?**

Of course, that table could go on infinitely (the equation ($x^{50,835} + 9 = 0$) is a perfectly legitimate polynomial equation), but as I mentioned in Chapter 3, the next type beyond quintic equations is usually just called "sixth-degree equations," then "seventh-degree equations," and so on.

(By the way, you may be somewhat surprised to see linear equations on that list, but it makes a lot of sense when you consider it. After all, you can think of the (**3x + 2**) in (**y = 3x + 2**) as a first-degree polynomial.)

At this point, thanks to the Quadratic Formula as well as to your other algebraic skills, you can solve any quadratic equation. As it happens, most equations past quadratics are beyond the scope of this book. Why should that be the case?

Well, interest in quadratic equations began, as far as we can tell, in ancient Egypt. Mathematicians there developed techniques for dealing with quadratic equations that essentially involved using tables to estimate the values of the solutions. About a thousand years later, in roughly 400 BCE, the Greeks and the Chinese also became interested in quadratic equations, but it was a Hindu mathematician named Brahmagupta from the subcontinent of India who, around 700 CE, first came up with a version of the Quadratic Formula very like the one we use today. Still, as far as we know, it wasn't until a European mathematician, doctor, and astrologer named Girolamo Cardano began studying cubic equations around 1545 that anyone did any serious work on a generalized method for solving higher-degree equations.

There is a Cubic Formula — I'll show it to you in a moment — and there are established algebraic methods for solving quartic and quintic equations, but in the nineteenth century a young Norwegian mathematician named Niels Abel proved that it is impossible to solve most equations of higher than fifth degree using ordinary algebraic methods.

So not only is it actually impossible (at least as part of an algebra class) to solve most higher-degree equations, but even most cubic, quartic, and quintic equations are beyond what you'll learn to do in this book. To get an idea of why, have a look at the Cubic Formula, used for solving cubic equations of the generic form $ax^3 + bx^2 + cx + d = 0$:

$$x = \sqrt[3]{\left(\frac{-b^3}{27a^3} + \frac{bc}{6a^2} - \frac{d}{2a}\right) + \sqrt{\left(\frac{-b^3}{27a^3} + \frac{bc}{6a^2} - \frac{d}{2a}\right)^2 + \left(\frac{c}{3a} - \frac{b^2}{9a^2}\right)^3}}$$
$$+ \sqrt[3]{\left(\frac{-b^3}{27a^3} + \frac{bc}{6a^2} - \frac{d}{2a}\right) - \sqrt{\left(\frac{-b^3}{27a^3} + \frac{bc}{6a^2} - \frac{d}{2a}\right)^2 + \left(\frac{c}{3a} - \frac{b^2}{9a^2}\right)^3}} - \frac{b}{3a}$$

Whoo! I will not ask you to try using that formula even once.

So where does that leave us? Well, there are certain higher-degree polynomials that are solvable using the techniques that you've already learned. Essentially, solving them means using variations on factoring and using the Quadratic Formula.

The first thing to realize is that the Zero-Product Property applies to any number of things being multiplied together. In other words, if $(a)(b)(c)(d)(e) = 0$, then at least one of those variables must be equal to zero.

The cubic equation $(x^3 - 4x^2 - 7x + 10 = 0)$ can be factored like this:

$x^3 - 4x^2 - 7x + 10 = 0$
$(x + 2)(x - 5)(x - 1) = 0$

2.   What are the three solutions to $(x^3 - 4x^2 - 7x + 10 = 0)$?

For each of the following factored equations, determine the degree of the equation and its solutions. (You can multiply each whole equation out in order to find its degree, but you don't have to; you only have to figure out what the highest exponent would end up being if you *did* multiply it out.)

3. $(x + 2)(x - 10)(x + 7) = 0$ 4. $(x - 7)(x + 1)(x + 3)(x - 2)(x + 9) = 0$

5. $(x + 6)(2x - 1)(3x + 2)(x + 5) = 0$ 6. $x(x + 5)(2x + 3) = 0$

7. $6x(x + 2)(3x - 5)(x + 1) = 0$ 8. $x^2(x - 5)(x + 2) = 0$

9.   Compare the degree of the equations in Problems 3 through 8 with the number of solutions to each problem. In all cases but one, what is true of the number of solutions and the degree of the equation? Based on this, are you willing to make a hypothesis about the degree of an equation and its maximum number of solutions? If so, what is that hypothesis?

If you hypothesized that the maximum number of solutions to an equation is equal to the degree of the equation, you were correct. It makes a lot of sense, really. After all, the factored form of a cubic equation, for example, can only have three parts — three "x's," if you want to think of it that way. The factored form of a cubic equation could be, say:

$$x(x - 4)(2x + 3) = 0$$

If there were any more x's in the factored form, it would be a higher-than-third-degree equation. Hence, a third-degree equation can have a maximum of three solutions, a fourth-degree equation four solutions, and so on. Notice, however, that this rule only gives us a *maximum*. For instance, the equation from Problem 8, ($x^2(x - 5)(x + 2) = 0$), is fourth-degree but has only three solutions.

It is possible to solve some higher-degree equations using the factoring techniques you know. In order to do so, you first convert the equation into standard form and then factor the left-hand side according to the techniques that you learned in the last lesson.

Find the solutions to the following equations and check to see that the solutions really do make the original equations true (unless your solutions are fractions; in that case you can check them with your math partners).

10. $x^3 - x^2 - 6x = 0$

11. $3x^3 - 9x^2 - 12x = 0$

12. $5x^3 + 20x = 20x^2$

13. $x^4 - 8x^3 + 15x^2 = 0$

14. $6x^3 + x^2 - 9x = -2x^2$

15. $3x^5 = 75x^3$

16. $x^4 - 5x^2 + 4 = 0$

17. $x^4 + 3x^2 - 10 = 0$ **(Although it looks similar to the last equation, this one has only two real solutions, and those solutions are irrational.)**

As I said earlier, it's possible to combine factoring with using the Quadratic Formula in order to solve some polynomial equations. For example, consider this equation:

$$3x^3 + 2x^2 - 2x = 0$$

18. **Begin factoring ($3x^3 + 2x^2 - 2x = 0$) by factoring out the greatest common factor of the terms of the polynomial.**

Your work in Problem 18, combined with the Zero-Product Property, means that the solutions to ($3x^3 + 2x^2 - 2x = 0$) are that ($x = 0$) and that ($3x^2 + 2x - 2 = 0$). You could try to factor ($3x^2 + 2x - 2$) in order to find the other two solutions, but I'll tell you right now that it isn't factorable.

19. Instead, use the Quadratic Formula on $(3x^2 + 2x - 2 = 0)$ to find the other two solutions to $(3x^3 + 2x^2 - 2x = 0)$ and then state all three solutions clearly.

Solve the following equations by combining factoring with the Quadratic Formula. (You can check your work with your math partners instead of plugging your solutions back into the equations.)

20. $8x^3 + 36x^2 - 4x = 0$

21. $15x^5 + 9x^4 = 6x^3$

Solve the following equations using whatever methods you prefer. (When the solutions do not contain fractions or square roots, you should check to see that they make the original equation true; if they contain fractions or square roots, you can check them with your math partners.)

22. $2x^3 + 2x^2 = 12x$ 23. $4x^4 = 6x^3 - x^2$

24. $x^4 - 13x^2 + 36 = 0$ 25. $x^2(x^2 + 1) + 4 = x(4 + x^3)$
(Use the Distributive Rule, rearrange it into standard form, and then factor it.)

26. $x^6 + 7x^3 - 8 = 0$ 27. $x^8 - 17x^4 + 16 = 0$
(Be sure to factor this one as far as possible.)

28. $12x^3 + 15x^2 + 9x = 0$ 29. $x^2(x^2 - 2) = 15$

30. Write a *Note to Self* about *solving higher-degree equations*. It should explain how to determine the maximum number of solutions that an equation can have, it should explain the combination of techniques you can use to solve higher-degree equations, and it should, of course, contain at least one well-explained example.

31. While I was creating problems for this lesson, I discovered an odd thing. Consider this equation:

$x^4 = x^2(5x + 6)$

Solve it by using the Distributive Rule, rearranging it into standard form, and then factoring.

Now go back and solve the equation again, but this time have your first step be dividing both sides by x^2.

What is strange about the two different sets of solutions that you just got to the same equation? Can you come up with any explanation for this situation? (In all honesty, I'm not sure that I can. I have a notion about it, but I'm definitely not sure.)

Solve the following sets of simultaneous equations:

1. 2x - 3y = 145
    -5x + 2y = -280

2. 20x + 4y = 0
    -55x + 3y = -7

3. Convert $\sqrt{28x^3y^{-2}}$ to simplest radical form.

4. Simplify the expression $\sqrt{75x} + \sqrt{12x}$.

5. Graph the following simultaneous inequalities:

    $y \leq \sqrt{x}$
    $y > -3$

6. Write $\dfrac{4x^2 + 3x - 2}{x^2}$ as the sum of an integer and a fraction.

7. Simplify the following expression:

    $$\dfrac{\dfrac{2x}{5} - 3}{x - \dfrac{15}{2}}$$

8. Solve the equation $3(m + 2) - 2(m - 3) = 5(m + 4) - 2(m + 2)$.

9. Linus and the armadillo are sorting through their comic book collection. They have three-fifths as many copies of *The Amazing X-Men* as copies of *Elfquest*. Their combined total of copies of *Cerebus* and of *Daredevil* is five times their number of copies of *Elfquest*, and they have 57 more copies of *Cerebus* than of *Daredevil*. Excluding *Elfquest*, they have 140 comic books. How many of each title do they have?

10. One last puzzle from Norman Willis (sort of). Definitely check out his book *False Logic Puzzles* if you enjoy these.

    Three boxes are laid out on a table. Two contain carcinogenic coal dust and one contains pixie stick dust. Here are the labels:

This is not the box to open unless the label on the adjacent box is true.	Exactly two of these labels are false.	This is the box to open unless the label on the adjacent box is false.

Which one are you going to open?

5 TWO-VARIABLE EQUATIONS OF HIGHER DEGREE

Higher-degree equations with two variables work in fundamentally the same way as quadratic equations with two variables: their graphs are curves (though, as you will see, of somewhat different shapes from those of quadratic equations); you can find the **x-intercepts** of their graphs by finding the points where **y** is equal to zero; and they are grouped in "families" with similar characteristics.

We'll look at cubic equations first. In the same way that all quadratic equations bear a fundamental resemblance to either $(y = x^2)$ or $(y = -x^2)$ — that is, the graphs of all quadratic equations are either upward-opening or downward-opening parabolas — all cubic equations are related to $(y = x^3)$ or $(y = -x^3)$.

1. Make a graph of $(y = x^3)$. You will need quite a bit of space for this: your x-axis should go from about -5 to 5 and your y-axis should go from -30 to 30. In order to do this properly, you'll need to tape two full-size sheets of graph paper together. When you're done, you can attach those sheets into your notebook as a fold-out. Graph the points where x = -3, -2, -1, 0, 1, 2, and 3. Connect those points with a smooth curve. Then, on the same set of axes but using a different-colored pen or pencil, graph $(y = x^2)$. Label each of the curves.

2. The graph of $(y = x^3)$ differs from the graph of $(y = x^2)$ in two significant ways. Compare the left-hand halves of the two curves (that is, the parts where x < 0). What is the first way in which the curves differ? Compare the right-hand halves of the curves (where x > 0). What is the second way in which the curves differ?

3. On the same set of axes but using a third color of pen or pencil, graph $(y = -x^3)$. Use the same x-values as you did for $(y = x^3)$. Label the curve.

Just as the graphs of quadratic equations have two different basic shapes:

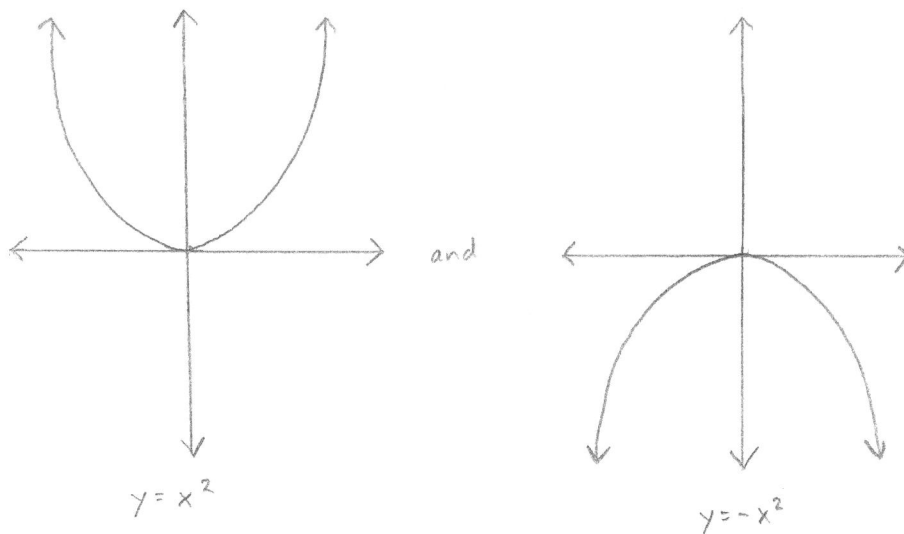

$y = x^2$ and $y = -x^2$

... the graphs of cubic equations have two different basic shapes:

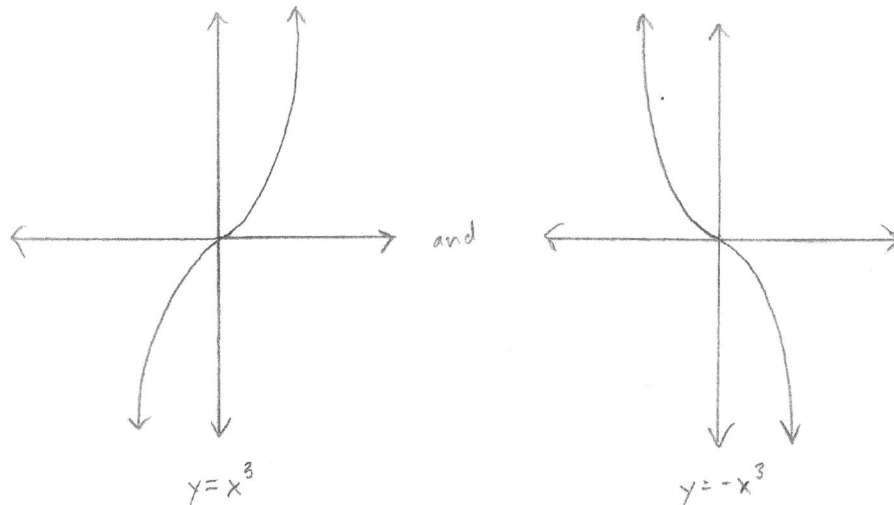

$$y = x^3 \qquad \text{and} \qquad y = -x^3$$

... but there is a lot of room for variation on those two basic shapes. In fact, I think it's fair to say that the graphs of cubic equations can vary more than the graphs of quadratic equations.

4. As you know, a one-variable cubic equation can have up to three real solutions. What does this fact imply about the graph of a two-variable cubic equation?

5. Factor the right-hand side of ($y = x^3 - 4x$) in order to find the x-intercepts of its graph.

6. Set up a pair of axes that go from ($x = -7$) to ($x = 7$) and from ($y = -16$) to ($y = 16$). Mark the three points from Problem 5 on the x-axis. (If you're working in a small graph-paper book, you'll probably need to make another fold-out page.)

7. In order to figure out the shape of the graph of ($y = x^3 - 4x$), plot the points where $x = -3, -1, 1,$ and 3. Connect all seven points with a smooth curve. Here are a couple of things to be aware of: the points where $x = -1$ and 1 are the vertices of the curve (unlike a parabola, it has two); and when $x < -2$ or $x > 2$, the curve should resemble ($y = x^3$).

8. Examine both your graph and the equation ($y = x^3 - 4x$). When $x < -2$, does x^3 push the graph up or pull it down? (To put it the proper way, is x^3 positive or negative?) Does $-4x$ push it up or pull it down? For those values, is x^3 "stronger" than $-4x$ or is $-4x$ "stronger" than x^3? ("Stronger" is in quotes because it's not really a technical, mathematical term — at least not in this context — but I hope you see what I mean.)

9. When $-2 < x < 0$, does x^3 push the graph up or pull it down? Does $-4x$ push the graph up or pull it down? Which one is stronger?

10. When $0 < x < 2$, does x^3 push the graph up or pull it down? Does $-4x$ push the graph up or pull it down? Which one is stronger?

11. When $x > 2$, does x^3 push the graph up or pull it down? Does $-4x$ push the graph up or pull it down? Which one is stronger?

12. Without plotting it out point-by-point, sketch the shape that you think ($y = x^3 + 4x$) would have. Think about your answers to Problems 8 through 11. Do $x^3 + 4x$ pull against each other? If so, at which points is which one stronger?

13. Based on your answer to Problem 12, how many real solutions does the single-variable equation ($x^3 + 4x = 0$) have? Why?

Factor the following equations in order to make sketches of their graphs. The only things you need to get absolutely right in those sketches are the x-intercepts of the graphs and their basic shapes.

14. $y = x^3 + x^2 - 6x$

15. $y = x^3 + 4x^2 - 5x$

16. $y = -x^3 - 6x^2 - 8x$ (I suggest using the technique where you factor out -1 from the right-hand side first. When you sketch the graph, ask yourself whether it resembles ($y = x^3$) or ($y = -x^3$).)

Sketch the graphs of these factored equations:

17. $y = (x + 3)(x - 2)(2x + 1)$

18. $y = -1(x - 5)(x +1)(x + 3)$

19. A single-variable quadratic equation can have as few as zero real solutions. What is the minimum number of real solutions for a single-variable cubic equation? Explain your answer.

20. Is it possible for the graph of a two-variable cubic equation to have two x-intercepts? If so, sketch the graph. Based on this, can a single-variable cubic equation have exactly two solutions?

Now let's look at quartic and quintic equations (and, indeed, all other higher-degree equations). I won't be asking you to make proper graphs of higher-degree equations the way you graphed ($y = x^3$). As you'll see in a moment, the fold-outs you'd need to make would get ridiculously big.

21. Copy and fill in the following table. Ask your teacher whether you can use a calculator.

x	x^2	x^3	x^4	x^5
-5				
-4				
-3				
-2				
-1				
0				
1				
2				
3				
4				
5				

22. Based on the table you just made, sketch the graphs of $(y = x^2)$, $(y = x^3)$, $(y = x^4)$ and $(y = x^5)$ on the same set of axes. By "sketch" here I mean that you shouldn't worry about particular points at all (except the origin), only about comparing the various graphs. I suggest devoting a big set of axes to this comparison and drawing each curve in a different color. Make sure to label the curves.

23. What would you expect the graph of $(y = x^6)$ to look like? You can either draw a sketch or describe it in words. What would you expect the graph of $(y = x^7)$ to look like? How about $(y = x^{10})$? $(y = x^{35})$?

24. A single-variable quartic equation can have up to four solutions. What does this fact imply about the graph of a two-variable quartic?

25. Consider this factored quartic equation:

$y = (x + 2)(x + 1)(x - 1)(x - 2)$

Sketch its graph, making sure only that you get the x-intercepts and the basic shape right. (If it helps, this graph will have three vertices.) Keep in mind how you expect the graph to behave when x < -2 and x > 2.

26. Now consider this factored quintic equation:

$y = (x + 2)(x + 1)(x)(x - 1)(x - 2)$

Notice that it crosses the x-axis in five places, which is the maximum for a quintic. Sketch its graph just as you did the graph for Problem 25, keeping in mind what the graph should look like when x < -2 and x > 2.

27. Based on the work you've done so far, what do you think the maximum number of vertices of the graph of a seventh-degree equation would be? What about the maximum number of vertices for the graph of a tenth-degree equation? What is the maximum number of vertices for the graph of an equation of degree n?

For all of the remaining problems in this lesson, if I ask you to make a sketch, it can be very rough — get the overall shape of the graph right and the number of times that the curve crosses the x-axis. Assume that each graph has its maximum number of vertices. If it isn't possible to make the kind of sketch that I ask for, explain why it isn't possible.

28. Sketch the graph of a two-variable quartic equation whose related single-variable equation has no real solutions.

29. Sketch the graph of a two-variable quartic equation whose related single-variable equation has one real solution.

30. Sketch the graph of a two-variable quartic equation whose related single-variable equation has two real solutions.

31. Sketch the graph of a two-variable quartic equation whose related single-variable equation has three real solutions.

32. Sketch the graph of a two-variable quintic equation whose related single-variable equation has no real solutions.

33. Sketch the graph of a two-variable quintic equation whose related single-variable equation has one real solution.

34. Sketch the graph of a two-variable quintic equation whose related single-variable equation has two real solutions.

35. Sketch the graph of a two-variable quintic equation whose related single-variable equation has three real solutions.

36. Sketch the graph of a two-variable quintic equation whose related single-variable equation has four real solutions.

37. Sketch the graph of a two-variable sixth-degree equation whose related single-variable equation has six real solutions.

38. Sketch the graph of a two-variable sixth-degree equation whose related single-variable equation has three real solutions.

39. Sketch the graph of a two-variable sixth-degree equation whose related single-variable equation has one real solution.

40. Sketch the graph of a two-variable sixth-degree equation whose related single-variable equation has no real solutions.

41. Sketch the graph of a two-variable seventh-degree equation whose related single-variable equation has seven real solutions.

42. Sketch the graph of a two-variable seventh-degree equation whose related single-variable equation has three real solutions.

43. Sketch the graph of a two-variable seventh-degree equation whose related single-variable equation has no real solutions.

44. Sketch the graph of a two-variable tenth-degree equation whose related single-variable equation has eight real solutions.

45. Sketch the graph of a two-variable thirteenth-degree equation whose related single-variable equation has ten real solutions.

46. In general, which sorts of equations can have no real solutions and which sorts always have at least one real solution?

47. Well, here it is. Your very last *Note to Self* for this series of textbooks. It should explain the art of *graphing higher-degree equations*. It should explain what different "families" (i.e. cubics, quartics, and so on) look like and what determines how many times a curve touches the x-axis. A few sketches seem like a good idea.

REVIEW

Complete the square to solve the following quadratic equations:

1. $x^2 - 6x + 2 = 0$

2. $x^2 + 6x = -6x - 28$

Complete the square in order to convert the following equations into vertex form:

3. $y = x^2 + 4x - 6$

4. $y = -x^2 + 10x - 19$

5. A sheet of metal 8 feet by 14 feet is being bent into a U-shape to make a shipping container. How tall should the sides be to maximize the volume?
(This is a tough one! Get help if you need it.)

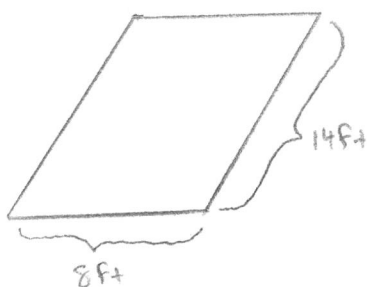

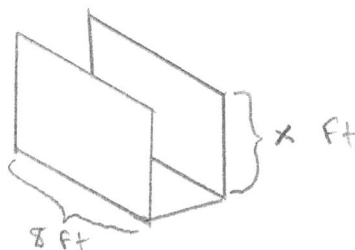

14ft
8ft
8 ft
x ft
not drawn to scale

Use the Quadratic Formula to solve the following equations:

6. $5x^2 + 3x - 1 = 0$

7. $5x^2 - 4x - 1 = 0$

Use the discriminant to tell how many real solutions each of the following equations has.

8. $10x^2 + 5x - 6 = 0$

9. $2x^2 - 4x + 2 = 0$

Factor the following polynomials:

10. $x^4 - 5x^3 - 84x^2$

11. $9x^5 - 100x$

Solve the following equations:

12. $x^3 + 5x^2 - 39x = -5x^2$

13. $2x^4 + 2x^3 = x^2 - x^4$

14. Find the x-intercepts and sketch the graph of the equation $y = x^3 - x^2 - 6x$.

15. One final puzzle:

Kubla Khan has decreed a strange ritual in his palace, Xanadu. There are 1,000 doors from the stately pleasure-dome into the gardens bright with sinuous rills and incense-bearing trees, and each morning the khan sends his 1,000 guards on their rounds in the following manner:

The first guard must go to every door and open it. The second guard must go to every second door and close it. The third guard goes to every third door and, if it is closed, he opens it, and if it is open, he closes it. The fourth guard does this to every fourth door, and so on. After the 1,000th guard has completed his duty, how many doors in Xanadu are open?

AFTERWORD

I've asked you to do many things over the course of these three books — to tackle tons of challenging problems, to do a lot of thinking, to do a lot of writing — and in this Afterword I'll ask you to do just two more things. The first one is to get out your Note to Self book and look through it. Just take a minute to appreciate what you've accomplished. I've never sat down and counted how many Notes to Self I've asked you to write, but my guess is that it's probably in the neighborhood of a hundred. Assuming you started at the beginning of *Jousting Armadillos*, your first Note to Self is about inductive reasoning and your last one is about graphing higher-degree equations. That means that those Notes cover everything from pre-algebra all the way through a traditional Algebra I course and at least part of a traditional Algebra II course. One way of thinking about those Notes to Self is that you've basically written a textbook yourself. And you've done a really incredible amount of work along the way. So the second thing I'll ask you to do is simply to pat yourself on the back. Congratulations.

So now what? Well, your career as a mathematician is not over. You will probably move on from here to study geometry. After that, many schools have slightly different course structures. Algebra II is something that's often offered, as is trigonometry. Then there is pre-calculus and calculus. Many schools offer a statistics class. If you decide to go on to study math in college, you may take classes called things like "Number Theory" or "Combinatorics" or "Real Analysis" or "Abstract Algebra." The possibilities for continuing study in mathematics are quite literally endless.

Of course, there's a good chance that you will only take a few of those classes. You may not become a professional mathematician, or an economist, or a statistician, or a math professor, or an engineer, and so your formal career in mathematics may end in just a few years. Why, you might ask yourself, have I bothered to get this far in math if I'm not going to use it in my profession? In answer to that, I would say that life and learning are about far more than the things that you will eventually do to earn a living. Chances are that you will not be a professional poet or a novelist. But that's no reason not to read poems and novels. Math — and, especially, I would argue, algebra — is just like poetry in that way. It makes your life richer even if you never earn a dime from it.

I've worked hard to try to help you understand that algebra is not just a powerful problem-solving tool (though it certainly is a powerful problem-solving tool), but a way to enhance and enrich your understanding of the world around you. Algebra can, as you know, tell stories — real stories. Those stories are important and useful and — yes! — they are also beautiful. For myself, I don't care too much whether you've memorized how to add and subtract fractions with variables or solve simultaneous equations or factor polynomials. After all, that's why you have that Note to Self book. But I do care very much that you have understood how those things work. Algebra isn't a set of skills to be learned, one after another, so that you can pass a test. Algebra, if you really understand it, is like poetry: a lens through which you can view the world.

I've referred to myself as "I" and "me" throughout these three books, and in a sense that's correct — it has been me, Linus, writing most of the actual words that you've read. But in another sense that "I" is really a "we." There is Greg Neps, of course, my co-author, who's been responsible for generating many of the problems that you've worked on. (That's right, you can blame him for some of them!) Then there's Sarah Pope, the editor of these books, who's a great deal more than an editor. She's spent uncountable hours working on every aspect of the books. She's responsible for a lot of the creative touches in them and she's also written a few of the problems. (You can blame her for the devilishly hard logic puzzle about knights at a jousting tournament in *Crocodiles & Coconuts*.) There's Mary Elliott, an Arbor School parent, who donated her time and expertise to develop and fine-tune the design of this series. There's Matt Dunkel, my predecessor as an algebra teacher at Arbor and the one who did the original work on which a lot of the material in these books has been based. There's Kit Abel Hawkins, the head of Arbor School. The whole project of writing these books was her idea, and without her incredible energy and vision, without her careful mentorship, they certainly wouldn't exist. There are other people as well, in particular the members of the Bloomfield Family Foundation, whose support — financial and practical and moral — have been vital over the several years that it's taken to get these books written. I can't even list all of those folks. And finally there are the students, at Arbor and at other schools, who have worked through these books, who've crafted them, who've spotted errors, who've demonstrated their own incredible work ethic and sense of humor time and time again throughout the process. And, after all, the students are what these books are all about.

You are what this is all about.

So, at last, thank you.

Linus
November 2011

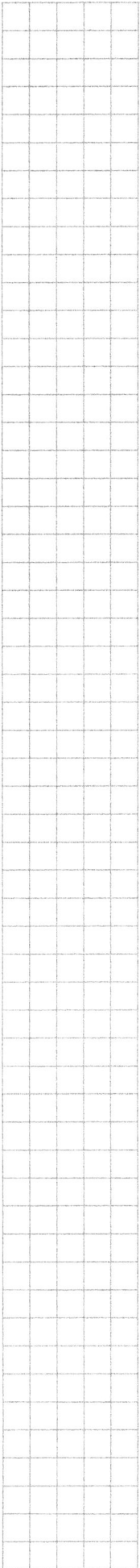